Approaching Chaos

Chaos

Can an Ancient Archetype
Save 21st Century Civilization?

First published by O Books, 2010
O Books is an imprint of John Hunt Publishing Ltd., The Bothy, Deershot Lodge, Park Lane, Ropley,
Hants, SO24 0BE, UK
office1@o-books.net
www.o-books.net

Distribution in:

UK and Europe
Orca Book Services
orders@orcabookservices.co.uk
Tel: 01202 665432 Fax: 01202 666219
Int. code (44)

USA and Canada
NBN
custserv@nbnbooks.com
Tel: 1 800 462 6420 Fax: 1 800 338 4550

Australia and New Zealand
Brumby Books
sales@brumbybooks.com.au
Tel: 61 3 9761 5535 Fax: 61 3 9761 7095

Far East (offices in Singapore, Thailand,
Hong Kong, Taiwan)
Pansing Distribution Pte Ltd
kemal@pansing.com
Tel: 65 6319 9939 Fax: 65 6462 5761

South Africa
Stephan Phillips (pty) Ltd
Email: orders@stephanphillips.com
Tel: 27 21 4489839 Telefax: 27 21 4479879

Text copyright Lucy Wyatt 2009
www.approachingchaos.co.uk

Design: Stuart Davies

ISBN: 978 1 84694 255 6

A CIP catalogue record for this book is available
from the British Library.

Cover design: Mortimer Design Ltd.

Printed by Digital Book Print

O Books operates a distinctive and ethical publishing philosophy in
all areas of its business, from its global network of authors to
production and worldwide distribution.

Approaching Chaos

Can an Ancient Archetype
Save 21st Century Civilization?

Lucy Wyatt

BOOKS

Winchester, UK
Washington, USA

CONTENTS

Prologue

I did not plan to write this book. When I first started researching material for a different book some years ago the future of planet Earth was not uppermost in my mind. At the time I was more concerned with exploring the origins of alternative medicines like homeopathy than anything to do with civilization. I was driven by a reluctance to agree with friends in the medical profession who believed that such treatments were 'unscientific' and therefore not worthy of consideration.

I disagreed, having found since the early 1990s that homeopathy worked for me, and yet I felt unable to refute the 'non-scientific' argument. I decided that, if I knew more about homeopathy's early development from the ancient Greeks onward, I would be able to formulate an effective response to my professional friends. As it happens, I did, but not in the way that I anticipated.

In the end I abandoned my original project because I began to realize that ancient history was of more than antiquarian interest. There was a whole world of ideas, both practical and philosophical, that could have extraordinary relevance for us today. Over several years I began to challenge my understanding of not only the development of medicine but also the accepted history of 'civilization'. In my intense exploration of the essence of civilization I began to uncover what could be described as a little understood 'formula' for civilization with which we had lost touch over time. But perhaps this could be reinstated – or at least aspects of it – to our benefit.

This process of questioning assumptions I had always taken for granted was deeply shocking to me. Nothing in my otherwise conventional educational background, with much deference to academia, prepared me for a profound reassessment. I had grown up in the university town of Cambridge in the East of

England where I felt the oppressive weight of serious academics past and present. This included generations of ancestors such as a particularly illustrious one, a Master of Queens, who in the 19th century had held the same Lucasian Chair in Mathematics as Sir Isaac Newton and Dr Stephen Hawking. My terror of academics was matched only by the fear of 'keep off' college grass signs ingrained in all Cambridge residents. But for some reason I did have the courage to question orthodoxy, having also inherited a somewhat rebellious nature. Indeed, when the time came, I deliberately chose not to apply to Cambridge University, and instead spent several happy years at Sussex University gaining a couple of degrees.

Two factors in my upbringing encouraged my willingness to question. First, I had the physical freedom to explore Cambridge and its cyclable environs in the 1960s and 1970s without parental constraint. This freedom perhaps gave me a greater sense of independence than might be the case today, especially for girl. Secondly, my parents, although of contrasting social backgrounds, had in common an unusual trait: surprisingly neither of them had been baptized at birth, and their own parents had been religiously tolerant – either Anabaptists, Quakers or free-thinkers – attached to no religion in particular. As a result, I was allowed to form my own opinions.

Nevertheless, even I have struggled emotionally with some of the conclusions that I have reached in the course of my research. I have not, however, been able to deny the evidence. I have tried to present what follows as logically and clearly as possibly. For, although the interpretations and ideas are my own, the evidence on which they are based is all from established published sources. I have not made up any of it – I haven't needed to, since the physical evidence alone is quite astonishing when looked at from a different angle.

I have been enormously lucky to have had the time – thanks to a supportive family – and the inclination to pursue such a fascinating subject area. I am especially grateful to my father Geoffrey Clarke for his work on the illlustrations, my publisher, editors, agent and all those loyal friends, too numerous to mention, who helped me make sense of my manuscript. I can only hope they still agree with me that a better understanding of the past could make all the difference to the way we cope in the future.

Lucy Wyatt
Suffolk, England, 2009.

Chapter 1

Civilization and 21st Century Chaos

Can civilization survive in the 21st century? By civilization I mean 'city life' and the comforts of a lifestyle associated with urban living – 'civilization' having at its root the Latin word civis ('citizen'). What future for our cities and our way of life? As our crises seem to multiply, many people are beginning to ask these questions. Prof Martin Rees is one of those who give modern civilization only a 50/50 chance, and James Lovelock of Gaia fame is equally doubtful.

We could take comfort from the fact that civilization has survived the most devastating of catastrophes since ancient times – the fall of Rome, the Black Death, intermittent warfare and revolutions... One only has to consider the reconstruction that took place after the end of the Second World War to know that civilization is capable of recovering from chaos. So, why should civilization be more at threat now than even in the darkest days of the last world war? What is it about civilization, anyway, that we consider to be so special that we fear the loss of it?

We have an unquestioning belief in 'progress'. We regard history, both ancient and modern, as a series of events that have brought us through an inexorable, Darwinian process from the Stone Age towards ultimate perfection: modern civilization based on reason rather than on the superstitions of the past. The last Ice Age belongs to our primitive past. We have images of cavemen wielding reindeer axes and of woolly mammoths roaming around, and are glad that we have moved on from there. We feel that the appearance of early urban civilizations marked a definitive break with our primitive Stone Age hunter-gatherer past, a sign of progress and an important step on the path to becoming the 'civilized' people that we are today.

Thus, the moment that the Mesopotamians built the first recognizable cities in what is now known as Iraq, about 6,000 years ago, was when we started to become 'civilized'. We can identify these sites as cities thanks to the archaeological evidence of specialist skills and political and religious power bases that form the infrastructure of a city, differentiating it from a small town or village – that is, the comprehensive *functions* of a city, not just market places. This major change to an urban existence was about more than a different basis for exchanging goods.

In addition to the organizing of people within a concentrated physical space, these ancient cities embodied the adoption of certain values. Implicit in the word 'civilization' was a way of living that raised us above our wilder, animalistic natures and the 'barbarity' of the hunter-gatherer. Civilization did this by emphasizing notions of order, straightness, justice, balance, weighing, measurement and accuracy, in distinction to the uncontrollable irregularity and organic arbitrariness of the chaos of nature. In essence, 'civilization' always was – and still is – about having a different kind of relationship with nature. Cities are artificial constructs within the landscape that are created with the skill of the architect and engineer.

Whereas the hunter-gatherer lives at the whim of nature, civilization relies on scientific and technical knowledge that deliberately interferes with nature for our own anthropocentric benefit, as well as protecting us from it. Even farming, which we think of as linked to nature, is a deliberate imposition by humankind on the countryside. We tame what we believe to be the unpredictability of nature, thinking that an orderly life is safer, more comfortable and, above all, more convenient.

It is the *convenience* of urban life that is maybe what we most value about civilization today and perhaps what we most fear losing. For the last 250 years or so, since the start of the Industrial Revolution, we have also been living in an accelerated technological growth spurt which has continually improved the ease of urban living. Indeed, we have so successfully conflated science and technology with being civilized that we would struggle to understand how civilization could exist without them. But, can we continue to take such progress for granted in the 21st century? What factors might halt that progress? What could we do that would make any difference?

The Challenges

One key factor that could interfere with the march of progress could be the economy. We yet have to see the full effects of the downturn in the global economy that began to bite in 2008. Even so, civilization has survived crises such as the Great Crash in 1929 and the Depression that followed. Terrorism, particularly suicide bombing, has been another important threat: if one were to listen to the rhetoric of groups such as al Qaeda, one could believe that the destruction of Western civilization was their stated goal. And yet, discounting the possibility of extensive nuclear warfare, the world has always experienced conflict somewhere without civilization entirely collapsing.

Climate change is arguably the most critical challenge facing us because of its global impact on our most basic needs for survival: food and water. The world is already suffering shortages, with population pressure exacerbating the problem. Even so, in 2000 the UN's Food and Agriculture Organization (FAO) was able to report confidently that the world's agricultural output could support a population of 8 billion people by 2030. What has changed since the year 2000, in terms of the carrying capacity of the Earth, is the unforeseen negative impact of climate change on harvest yields. Global yields in the major staples such as wheat, maize and rice have been falling in recent years as harvests have suffered from the extremes of intense drought or flood. Whatever the primary cause, whether or not man-made, extreme weather is a reality we cannot ignore. The tsunami in Thailand, Hurricane Katrina in America and the typhoon in Burma were all dramatic indicators of the destructive power of nature.

Equally daunting are the issues surrounding water. Approximately a third of the global population now experiences water stress with either difficult access to safe water or an absence of water at certain times of the year. Whereas in the past people could rely on glaciers to compensate for lack of rain, this supply is also becoming unreliable as the glaciers retreat in a warmer world [Stern, 2007, p63]. Increasing numbers of rivers now dry up before they reach the sea. Likewise, deep and ancient aquifers are disappearing through overuse. As parts of the globe become uninhabitable, significant numbers of people will become dislocated, potentially exacerbating tensions that already exist within society. We cannot delude ourselves in the West that we are somehow protected from these tensions.

The West is not Immune

Food supply is increasingly becoming an issue even for the West. Britain could be the first of the Western democracies to experience social problems due to food shortages, especially given that it has one of the denser populations on Earth and depends on other countries for at least 40 percent of its food supply. Even without the effects of climate change, we in the West are not in a good position to respond to our food needs. Modern farming relies on an industrial model of production that is not sustainable, either environmentally or financially; neither can we rely on science to resolve the problems of improving yield, when yields are decreasing due to the long term effects of that very industrialization. Those concerned with a scientific solution to farming problems tend to overlook the sector's heavy depen-dency on hydrocarbons for fertilizer and fuel – and genetically modified crops rely on both.

Whatever happens to the price of oil, the era of cheap fuel is over. The bubble of growth based on easy exploitation of fossil fuels, beginning in the 18th century with the Industrial Revolution, is ending. We now face an uncertain future with regard to power supplies.

There is a further threat to power supplies which is quite separate from fossil fuel availability. This threat could even be the real driver of climate change. What we may be experiencing are the effects of a gradual alteration in the Earth's magnetic shield, potentially leading to a future geomagnetic pole reversal whereby north would become south. According to Prof David Gubbins, from the School of Earth Sciences at Leeds University, activity in part of the Earth's core suggests that 'this could be the start of a reversal. A weaker magnetic field means more cosmic radiation reaching Earth's surface because of the lessened shielding effect of the magnetosphere' [*Nature* 452, pp165-167].

In addition, our weather systems would become increasingly extreme in their patterns as they become more exposed to solar flare activity. More critically for modern urban life, as Prof Gubbins points out, any alteration in the magnetosphere resulting in an 'increased atmospheric geomagnetic activity means more disruption to electronic communication and power distribution' [*Nature* 452, pp165-167]. The consequences of a magnetic pole reversal would be chaos as magnetic north disappears, leading to potentially serious health hazards for human populations because the magnetic shield which currently protects us against harmful radiation from outer space would be undermined while the shift takes place.

Collapse

Of all the factors outlined above, it is, however, highly unlikely that any single one would destroy civilization in the 21st century. But how would we cope if these and other crises coalesced: crises with which we can cope on an individual basis, but which together could undermine the fabric of our societies? What would be the tipping point for a collapse in the 21st century?

Before some of these threats materialize, there is the possibility that our fear of chaos (and our fear of fear) will justify an
increasingly authoritarian tendency toward political and social
control, even in the liberal democracies of the West. We risk
falling for the political temptation to circumscribe our freedom
and our ability to determine our own futures. The apparatus of
government now has unprecedented possibilities of surveillance
and data collection, rendering us more vulnerable than ever to
arbitrary government.

How 'civilized' would life then be if we were discouraged
from deciding our own responses to crises, because 'state control'
took precedence? What if, in the middle of a severe economic
recession, we could no longer rely on the lights coming on when
we flicked a switch, and, at the same time, food availability
became an issue even in Western cities? Of course people would
adapt, but how well would they adapt if, on a regular basis, they
had to face some of the destructive changes that are occurring in
the physical world at the same time – the floods, the hurricanes,
the fires? *And if there was nowhere to go because nowhere in the world
was immune from these changes?* Without question, we have been
ill-prepared for what could happen, however much we might
cocoon ourselves.

Even in the highly sophisticated urban West, we are still
surrounded by ignorance, poverty, violence and prejudice,
coexisting in urban spaces that can depress the spirit more than
raise it. Investment in infrastructure – whether transport systems
or healthcare services – invariably fails to keep pace with expectations and soon deteriorates. Because cities have lost much of
their function as industrial centers, unemployment means that
people can feel denied a purpose in life. We have created soulless
places from which people understandably find diverse ways to
escape, whether through consumerism, religious extremism or
dependency on alcohol and drugs. People are neither happier nor

healthier, and not necessarily more civilized in their behavior than they were in the past.

Solutions

So, what should we do? Panic? No doubt many will. Either we change, or change could be forced upon us. Some will turn to God for answers, and some to science. God, on the one hand, has not always been much help with the catastrophes that have faced us throughout history. Science, on the other hand, has lots of ideas about high-tech solutions: genetically modified crops, biotechnology, cold fusion, nuclear power, nanotechnology... We are promised that Big Science is on the case. Because science and technology appear to have successfully gained control of the natural world, we are in danger of becoming arrogant about solving our problems.

But clever scientific answers are unlikely, at best, to have more than a limited impact on some of the challenges facing modern civilization. They cannot solve a more fundamental problem, which is that of *mindset*. Unless we change our way of thinking, no amount of hydrocarbon substitutes will mitigate the social effects of climate change. Developing a renewable energy strategy will be like sticking a plaster over an old wound, if we don't change the way that we think about how we live.

We have failed to realize that our obsession with progress has created an emphasis on goal-oriented thinking. Whether we admit it or not, we all tend to subscribe to the belief that 'the end justifies the means'. In the case of science, this philosophical construct is apparent when applied to the use of animals in laboratory experiments. In religion, it seems to be a characteristic of monotheism that many millions have died and continue to die, pointlessly, in the name of God. But does someone's interpretation of the will of Allah and the desire to go to heaven absolve an individual from all responsibility and justify blowing up innocent people? Did the good of your eternal soul justify the

7

Catholic Inquisition torturing you on the rack?

The effect of goal-oriented thinking generally on society is a short-term outlook on life that leaves us blind to the needs of nature. We put our own expediency, our need for bigger, better, faster, cheaper and easier, ahead of anything else. We sanction the most appalling environmental destruction because building an extra airport runway or a motorway bypass will bring us economic growth, which is more important than anything. The outcome has been a general lack of concern for the spirit of place and an acceptance of the desecration of our natural environment, as well as a willingness to allow urban life to damage its own life-support system.

Westerners in particular have a responsibility to find solutions. It is the value system of the West that has driven the extensive exploitation of natural resources since the start of large-scale fossil fuel use in the Industrial Revolution, encouraging the belief that the Earth exists for our immediate gratification and that we have no responsibility for it.

We have been warned for a long time that we were on a path to self-destruction. Native peoples in North and South America have been trying to tell us that we have not been listening to Mother Earth. As far back as the 19th century the great Sioux chief Tatanka Yotanka, or Sitting Bull (1831-1890), is attributed with saying, rather prophetically, 'Only when the last tree has died and the last river been poisoned and the last fish been caught will we realize that we cannot eat money'.

Our desire to be rational, scientifically-minded human beings also discourages us from involving *feeling* in our decision-making, thus implicitly justifying our insensitivity toward nature. Today the adjective 'scientific' is restricted to a narrow meaning concerning a specific method rather than any general sense of knowledge ('science' having at its root the Latin *scio*, 'I know'). We regard something as scientific only if it can be proven using an experimental method that produces the same results every time, regardless of who conducts the experiment. The results are then deemed to be 'objective', based on reason and not feeling.

Some point to the French philosopher Descartes as being the one who identified this separation of subjective and objective, between mind and body, psychological and physical. In his 1637 *Discourse on the Method* he wrote,

> From that I knew that I was a substance, the whole essence or nature of which is to think, and that for its existence there is no need of any place, nor does it depend on any material thing; so that this 'me', that is to say, the soul by which I am what I am, is entirely distinct from the body, and is even more easy to know than is the latter; and even if body were not, the soul would not cease to be what it is.

In distinguishing between the thinking thing, the *res cognita* and the non-thinking body, the mechanical *res extensa*, Descartes created the conditions for a moral viewpoint to arise: not only *are* feelings separate from the body, they *should* be [Damasio, 1996, p249].

While Descartes may have been right – that the body and the soul are distinct from each other – he may have unwittingly helped to lead Western philosophy into a dead end by discouraging us from valuing feelings when rationally interacting with the world. And so we lost our sensitivity to signals that our way of living had no long term future. The point that we all need to understand is that 'the end' does not automatically justify 'the means' – how we do something can be just as important as why.

To quote another 19th century Native American, Chief Seattle, in a famous speech of 1854, 'This we know: the Earth does not belong to man; man belongs to the Earth. All things are connected like the blood that unites us all. Man did not weave the web of life, he is merely a strand in it. Whatever he does to the web, he does to himself'. Only now, more than 150 years later, are we finding out the truth of this wise observation – everything is interconnected.

An Alternative

Ideally, we want a way of living that is more in harmony with each other and the natural world. We want a lifestyle that balances the comfort and convenience of an organized, civilized society living in cities, but with greater sensitivity toward the natural environment, its resources and energy supplies. For that to happen, we need a different belief system – one that gives us a framework for action. It is more important than ever that we rediscover a spirituality and practical knowledge base that makes new sense of the world and connects all parts to the whole. At the moment the debate around belief and knowledge is in danger of being polarized between 'irrational' religion and 'rational' science.

Maybe we are also about to discover that there is no inevitability to progress. Darwinian evolution does not necessarily apply equally to social change as it does to the development of species. History, both ancient and modern, has many examples of societies that have failed or declined. If we want to avoid becoming one of them, we now have an opportunity to reconsider fundamentally the meaning of the term 'civilized'. The idea of rolling back thousands of years of progress and returning to live in the equivalent of a Stone Age cave is just unthinkable and terrifying. We firmly believe that being 'civilized' and living in cities has so obviously been the right thing for humanity that we cannot contemplate any alternative.

Even so, there is much to be said for having another look at the ancient past. I am not suggesting a romantic notion of returning to a golden age. I would not argue that life was better in the ancient past – ancient societies had their problems. But not everything about the ancient past was necessarily backward. We could learn much about surviving from societies which we have been too quick to dismiss as irrational, primitive and pagan.

But when we do think about the ancient past, we tend to limit ourselves to admiring ancient Greek and Roman societies. It is the powerful Graeco-Roman mindset that continues to affect us and how we think about civilization, as well as our ideas about religion and science. Indeed, it is thanks to these Indo-European ancestors that we have inherited our semantic understanding of what civilization means. Their concepts relating to civilization are embedded in our language and thought-processes – although some of these concepts predate even the Greeks and Romans and are part of our very ancient Indo-European heritage going back at least 5,000 years.

If we really want to change our thinking, we need to question our attachment to our Greco-Roman heritage. There is another way of understanding civilization which is much older than the Classical era – one which first appeared with the earliest cities of Sumer over 5,000 years ago – and which was subtly different. It is these more ancient ideas which have such potential for us.

The Potential of the Past

No one should be under any illusion that people who lived over 2,000 years ago were in any sense primitive or inferior to us. They were capable of achieving as much as, if not more than, us. Monumental buildings and extraordinary feats of engineering – the pyramids and temples of the Middle East, Egypt and South America – are some of the most remarkable features of civilization.

It is easy to forget that these impressive monuments were built in the Bronze Age, some time in the third millennium BC, *before* the Iron Age, at a time when copper was the main material available for tool-making. But, copper, even when hardened, is useless for cutting stone, especially hard stone like granite. Yet the ancients were able to work with the hardest of stone. They had techniques for cutting stones to an incredibly high standard of precision. They made their decisions regardless of how far the building site might be from the source of materials. The Egyptians, for instance, used granite that they quarried from Aswan over 500 miles away.

All this was done for a reason; they did not build their monuments in order to impress us; they created these vast monuments because they wanted to and because they could. If we want a world where massive projects can be achieved without the use of hydrocarbons we have only to look out of the window of the Hotel Mena at Giza in Egypt. Staring us in the face is one of the wonders of the world, the Great Pyramid.

The Great Pyramid illustrates both the ancients' ability to achieve incredible precision and, at the same time, to build on an enormous scale. This single creation is perhaps the most awesome human achievement of all times. Its exact dating is somewhat controversial, although there is apparently relatively recent carbon-dating which puts it around 2850 BC, 400 years earlier than the more traditional date which is based partly on the pharaoh for whom it was allegedly built [Picknett & Prince, 1999, p43].

The statistics of the Great Pyramid are breathtaking. It covers an area of 13 acres and consists of approximately 2,300,000 blocks with an average weight of 2½ tons each, placed on solid rock. Some individual blocks are as much as 15 tons and the granite slabs roofing the King's Chamber are 50 tons [Hawkes, 1973, p347]. It was also once cased in 144,000 16-ton blocks of white polished Tura limestone, arranged in 216 courses from top to bottom and cut to fit with a perfect slope and with such precision that the joints were almost invisible to the naked eye. The entire pyramid, however, would have appeared intact from the outside because the original entrance was secretly located on the nineteenth course of the northern face [Alford, 2003, pp42].

What is most unnerving is its precision. Its four corners are true 90 degree angles 'to within 1/100th of an inch'; it is aligned on the cardinal points to the extent that it deviates by only 5 degrees; its sides are not equal in length, but the difference is only a matter of 7.8 inches; and the angle of its slope, 51 degrees, is known as the 'perfect angle' because it embodies the mathematical ratios of *pi* and *phi* (π and ϕ). Thus, the pyramid is said to have 'squared the circle' [Alford, 2003, pp41-43]. The Great Pyramid may be impressive, but it is not alone, particularly in the use of enormous blocks of solid stone.

On the Giza plateau there are several temples – the Sphinx Temple, the Valley Temple and the Upper Temple – which also contain massive blocks, one of which is estimated to weigh an incredible 468 tons. These temples are similar in style and construction to the Osireion, 270 miles south at Abydos [Collins, 1998, pp11-12]. At the temple of Serapeum at Saqqara there are 21 granite boxes in rock tunnels, each one with perfectly fitting lids and weighing 100 tons. The granite stone originally came from the Aswan quarries and each box is a complete section of solid granite – which would be impossible, given their size and weight, for a modern supplier who would need to supply them in pieces and then bolt them together [Hatcher Childress, 2000, p274, p276].

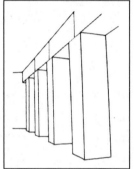

Columns in the Valley Temple, Giza

Besides Egypt, we find the same amazing use of cyclopean stone elsewhere. The 2,500ft long Temple of Ba'al-Astarte at Baalbek, east of Beirut, was situated on a platform made of gigantic stone blocks that became the base for the now ruined Roman Temple of Jupiter. Some of the individual stones possibly weigh 750 tons each and stand more than 20ft high. The largest block, however (70ft long, weighing 1,000 tons) was left behind in the quarry half a mile away. There was no discernible road between the temple and the quarry and it was not possible, given the terrain, to use rollers [Hatcher Childress, 2000, p41-p43, p45].

The same kind of evidence exists in Central and South America. In Bolivia, Tiahuanaco has 100-ton foundation blocks [Hatcher Childress, 2000, p55]. In Peru, the Sacsayhuaman fortress above Cuzco has perfectly cut massive limestone blocks

Megalithic blocks at Sacsayhuaman

over 100 tons each and built without any mortar – cut so perfectly that it is not possible to insert a knife blade between them [Morrison, 1978, p107; Hatcher Childress, 2000, p51]. The blocks for Cuzco itself were thought to have come from Ecuador, 1,500 miles away [Hatcher Childress, 2000, p52]. Some blocks in Puma Punku weigh 100-131 tons; others are andesite, the largest of which weighs 16 tons and is believed to have come from over 35 miles away [Morrison, 1978, pp141-143]. The 200-ton blocks of porphyry for Ollantaytambo and Ollantayparubo are believed to have been carried over the mountains and ravines of Peru before being placed on 1,500 foot high cliffs [Berlitz, 1975, p144].

Throughout the Mediterranean and Western Europe there are megaliths. On Malta there is a 5,000 year-old complex of cyclopean blocks at Tarxien, some weighing 50 tons [Rudgley, 1998, p23]. Dating also from fourth millennium BC there are 40-50,000 stone monuments, some weighing in excess of 50 tons, mostly in the coastal areas of western Europe – the most famous of which are at Stonehenge and Carnac in Britanny [van Doren Stern, 1969, p247].

The physical remains of these once great monuments cannot

but make us wonder how and why. Historian Philip van Doren Stern has calculated that it must have taken 400 years and '1.5 million man-days of exceeding hard labor' to build the first three phases of Stonehenge. As he says, 'How were the sparsely fed, poorly housed, ignorant and completely illiterate people of southern England persuaded to work so strenuously for so many generations?' [van Doren Stern, 1969, p262-3] – not least because this work was carried out without capstans, pulleys or scaffolding. Egyptians did 'not even employ wheeled carts before New Kingdom times' [Hawkes, 1973, p347].

An Ancient Model

In later chapters I shall explore how they were able to build on such a scale. Before that I want to examine the background to the civilization that lay behind such impressive monuments and understand better how civilization came about. In order to do that I shall go back to the time when the earliest hints of civilization began to appear – the earliest farming, pottery and metalwork – as far back as the end of the last Ice Age. By going so far back it becomes apparent, in particular by using recent archaeological information and modern techniques of enquiry such as genetic analysis, that there are strange anomalies surrounding the arrival of agriculture and cities. The ancients were even able to manipulate the power of nature for their own benefit in ways that invite comparison with the modern use of quantum physics.

What emerges from close examination of the past is an archetype of civilization that conforms to specific principles, to the extent that one can reconstruct a model of it. These principles were part of a common culture that existed across the continents, and covered every aspect of civilized life, establishing, above all, that cities could live in harmony with nature. There was no disconnect in this blueprint, no separation between science and religion. The Egyptians referred to it as 'living in truth' ('in

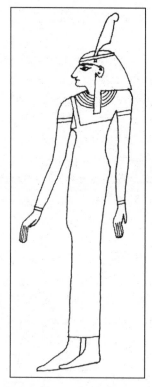

Egyptian goddess Ma'at wearing the feather of truth

ma'at'). Ancient Egypt became known as the best example of this ancient model, predating Greece and Rome, although it was not the only one. Evidence for the same concept exists in Mesopotamia, the Indus valley, the Phoenician coast, Minoan Crete and in Central and South America. The same model was able to survive various calamities over the millennia and enabled civilization to continue for a long time, especially in Egypt.

Civilization in Egypt began to decline and finally came to an end in the 4th century AD about the time when Rome, the dominant superpower in the region, adopted Christianity. Without the Romans taking an interest in this minor cult 200 years after it first appeared, it is possible that Christianity would have remained an obscure development from 1st century Palestine. This Romano-Christian combination was not quite the force for progress that we have been led to believe. Rome actively encouraged Christianity to destroy the last vestiges of the ancient model.

It is not without irony that it was from within another monotheistic religion, Islam, that the same principles re-emerged. For several centuries the Islamic world experienced a flowering of the ancient model, from Central Asia to Moorish southern Spain. Since the 4th century AD there have been several attempts to introduce parts of the model to the West. The first, to a limited extent, was around the time of the medieval Crusades; then, more noticeably, at the time of the Renaissance, the model reappeared through a body of writings called the *Hermetica* or the *Corpus Hermeticum*. It could be argued that the *Hermetica* was even the main inspiration for the changes associated with the Renaissance. Although the *Hermetica* was later discredited in the 17th century, there was something of a revival of interest in the academic study of the *Hermetica* in the early 20th century – coincidentally at about the same time that scientists chanced upon their discoveries in quantum physics. In spite of this coincidence of timing, few have realized the connections between modern physics, the *Hermetica* and other ancient philosophies.

If we want civilization to survive, now is the time for us to consider re-connecting modern life with our ancient past by re-incorporating certain aspects of the ancient model. If civilization were just the product of evolution then there would be little reason to re-examine ancient history, since we would presumably already be living in the most evolved aspect of it. The model of civilization that appeared so long ago exists as an archetype for *all* times and is therefore still available to us today.

Chapter 2

The *Ur*-Concept – the original model for civilization

Few of us give much thought to our prehistory. For many, especially in the West, civilization takes off properly only with ancient Greece and Rome. We might condescendingly chalk up Mesopotamia on the list of humanity's firsts – the first cities, first writing and first farming. We then lose interest. Their writing was in incomprehensible cuneiform, understood only by antiquarian scholars, and farming was an idea that was bound to happen anyway. Ancient Egypt is equally impenetrable with its hieroglyphics, strange funerary arrangements and the apparently pointless building of enormous pyramids. Similar pyramids appear in Central and South America but, in spite of their spectacular monuments, these civilizations are disregarded as being barbaric because they conducted human sacrifices. For us the remains of Mesopotamia, ancient Egypt and the Americas are seen as just so much exotic archaeology with little direct relevance to us, beyond providing a scenic backdrop for biblical stories in the Middle East.

We take an interest again with the ancient Greeks and Romans – partly because we still learn their languages. We are grateful to the Greeks for philosophy and geometry – histories of medicine and mathematics usually start with Hippocrates, Galen, Euclid and Pythagoras – and to the Romans for organizational improvements like law, government and straight roads. With the Classical era regarded as a high point of civilization, we then have the impression that life temporarily went wrong with the collapse of the Roman empire, and even call the era that follows the 'Dark Ages'. The few glimmers of light in this period, we are told, include the Arabs' invention of the zero, thus enabling mathematics to develop properly; and the growth of the Christian church. At last something familiar...

We are encouraged to think of the arrival on the scene of Christians as those who win through to save us from misguided pagan beliefs. Even the way that we have written our dating system (Before Christ and Anno Domini) is indicative of what makes us feel comfortable about the past. In other words, ancient Greece and Rome are important to us because they have shown us how to live and think – in terms of political philosophy, law, medicine or architecture, etc. But the Romans' conversion to Christianity, with its redemptive and salvationist message, we are taught to regard as the truly civilizing gloss.

Christian monotheism put Roman pantheism in its proper place as devil-worship. It meant that we no longer needed to rely on pagan superstition to explain the world. Instead we could take the ideas of the Greeks and concentrate on creating an objective, logical, verifiable approach to science, separate from our religious beliefs. Geniuses like Sir Isaac Newton and others in the Age of Enlightenment and the Industrial Revolution could then lay the foundation for our modern world. In particular, this ability to be 'scientific' today is what distinguishes us from our past and from what we still consider to be 'primitive' societies, which didn't even have the wheel, let alone writing.

Yet these primitive societies were not only capable of building amazing structures but also they lived highly complex urban lives thousands of years ago, creating the basis for the later Greek and Roman versions. What is more remarkable is that examples of civilization pre-dating the Greeks and Romans, in whatever timeframe and on whatever continent, all conform to certain patterns, to an archetype which I will refer to as the *ur-concept*.

I call it the *ur*-concept for reasons that will become obvious (*ur* being a word associated with 'foundation' or 'origin'). What follows is my interpretation of historical and archaeological evidence, unrestricted either by chronology or location, in order to identify the key elements of the archetype. Chronology is important, but it increases the likelihood of being distracted by our cultural preference for developmental progress because we like to presume a pattern of cause and effect. As a result, we have been looking at our ancient past through the wrong end of the telescope. This chapter is thus a cross section of millennia and continents in an attempt to reconstruct the archetype.

In general terms, this archetype is based on principles of order, organization, accuracy, measurement, rules, law, mathematics, geometry, and so on – all with the objective of creating a more convenient lifestyle. These principles are apparent in the following areas which distinguish urban societies from hunter-gatherers:

- Agriculture (in particular the domestication of animals and a highly organized plant production);
- Power – political power, with its physical manifestations of administration, architecture and infrastructure, and religious power, with the defined roles of the priesthood;
- All forms of communication (such as trade, travel and education);
- The individual body, more systematically cared for through the development of the art of cooking, and the link between food and medicine.

We shall now examine the *ur*-concept more fully in several ways, starting with domestic animals.

Animals

Until relatively recently, domestic animals have been an energy source for us: the means by which we have been able to plough fields or travel long distances overland. They also provide us with a source of protein and calcium. Civilization could not have happened without the transformation of wild animals into domesticated versions. We entirely take for granted the usefulness and docility of domestic animals, and it is difficult for us to envisage a time when it was not so. I will discuss in the next chapter the circumstances in which this change came about.

It is no coincidence that the kind of farming which is familiar to us today only really developed at the same time as cities began to appear in the fifth millennium BC. By 3600 BC Uruk in Mesopotamia, for instance, had wheeled transport. Before that, only a limited form of proto-farming existed involving domesticated cereals and wild animals.

Take the horse as one example; this animal has had a complex relationship with humans since domestication, the earliest evidence of which dates from the fifth millennium BC in the Ukraine. The Indo-Europeans were possibly among the first to benefit from the use of the horse because they lived in an area that was the natural region for the wild horse. On the one hand, the horse has been a means of traction and transport – enabling the deep steppe and remoter regions to be opened up to a pastoral way of life. On the other hand, the horse also became an instrument of political policy, an effective part of a war machine, used either for mounted warriors or charioteers, and in that context it connects with the exercise of political power.

Government
Civilization requires a king, a ruler, responsible for the implementation of right, of law, of fair administration. The ruler also had a duty of care towards his people which resulted in his participation in certain rituals, supervised by the priesthood, that were an essential part of the whole civilization concept.

The ideal of order as administered by the ruler is reflected in the physical design of a city. Three characteristics identify cities as conforming to the archetype: a pre-planned urban layout based on patterns; the perfect straightness of individual construction works; and the creation of massive engineering feats, often in the management of water supplies and other infra-structure projects.

In terms of urban design, there were three basic patterns – quartering, gridiron and circular. Examples of quartering are to be found in Mesopotamia and South America. The *Epic of Gilgamesh* refers to the division of Uruk into a square mile of city, a square mile of palm groves, a square mile of brick pits and the temple complex of the goddess Ishtar-Inanna, the remains of which have been found dating to the fourth millennium BC [Rohl, 1998, p172]. In South America, Tenochtitlàn, the original Mexico City, was divided into four districts [Allen, 1998, p78].

Other cities, like Akhetaten in Egypt and Mohenjo-daro in the Indus valley, followed more of a gridiron pattern. Akhetaten had three 30-47m (90-140ft) wide main streets running east-west and three north-south, paved with white stone [Feather, 1999, p147]. Similarly at Mohenjo-daro there were three 10m (30ft) wide streets running north-south and two crossing at right angles, creating twelve blocks which were subdivided by alleyways. The same north-south axis was repeated in the Indus valley at Harappa, Kalibangan, Chanhu-daro and Lothal [Hawkes, 1973, pp271-272].

The Phoenician city Byblos, or Gebel, on the Lebanese coast, had streets running in a concentric pattern around its temple – dedicated to Baalat-Gebal the Queen of Heaven – together with a canal system for rain collection and drainage [Knight & Lomas, 2000, p158]. The city was protected by a wall with one entrance to the land and one to the sea. Knossos on Minoan Crete was another place where the temple was at the centre of the city both geographically and economically. Surrounding the temple were big storerooms for agricultural produce and craft workshops where, among other things, silver, tin, copper and gold were worked [Castleden, 1993, p109].

Temples were not only sometimes literally but also symbolically at the centre of city life. Archaeologist Jacquetta Hawkes notes that the existence of the temple and its priestly caste distinguishes a city from being a mere central trading point. She cites the view of Robert Adam that 'Apparently the first specialized adminis- trative group, or ruling stratum,

Temple of Apollo, Dephi, Greece

was composed of hierarchies of priests' – although she is not sure why this should be the case [Hawkes, 1973, p8]. In fact, so important was the temple on Knossos that archaeologists have been puzzled by the lack of an administrative and governmental palace. Knossos, however, was not alone in lacking an adminis- trative centre: the same also applied to the Indus valley and other early cities in Mesopotamia. Sometimes they were located in a higher area than the rest of the city, raised up on a mound, a simple earth mound, or on a pyramid or ziggurat.

For the ancient Egyptians, the surveying and setting out of temples and other buildings was considered so important that they attributed the role of 'mistress of measure' to Seshat or Sefhet, the female counterpart of the god Thoth, one of the most influential of the whole pantheon of Egyptian deities. Sefhet always accompanied the foundation ceremonies of temples, carrying 'the measuring cord, a knotted rope that is divided into

Sefhet, Egyptian mistress of measure

segments of the sacred cube' [West, 1993, p48]. Sefhet was responsible for the dedication of the sacred precinct by means of the
'stretching of the cord'. She can be recognized by the seven-
pointed star or petalled flower on her head which '...is believed
to represent an ancient astronomical siting device' and emphasizes 'the fundamental criteria in the planning of the Divine
House – its association with sky phenomena'. The corners of a
new temple would be marked out at night using alignments on
an astronomical point. In fulfilling this function, Sefhet was
acting as an extension of Thoth in helping with 'recording time,
dispensing measures, defining space and presiding over the
apportionment of land to the *Neteru*', the gods [Clark, 2000,
pp213-214].

Whatever the project – a domestic palace, temple or aqueduct
– straightness was consistently a key characteristic, regardless of
the material used, whether bricks or stone, or the size of it
(individual blocks could well be several tons in weight). The use
of the plumbline or bob was fundamental to this aspect of the
model – the third millennium BC Mesopotamian *Epic of Gilgamesh*
makes specific reference to the great wall surrounding the city of
Uruk being as 'straight as an architect's string' [Rohl, 1998, p172].
Examples of plumb bobs can be seen in museums in Luxor,
Egypt, and in Delhi where Indus valley artifacts are on display.

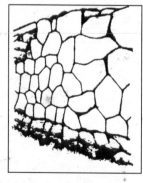

The exception to this principle was
the deliberate choice to use polygonal-
shaped blocks in certain building
works. The reason for this departure
from the otherwise regular pattern of
construction was that polygonal blocks
withstand better the effects of earthquakes. Examples of this style of
building can be found in the Americas,
the Mediterranean and the Middle East.
In spite of the irregular shapes, each one

Example of polygonal
blocks at Delphi, Greece

was still cut perfectly to fit like a jigsaw with its neighbors.

Water

The ancients were capable of building other huge-scale infra-structure works, such as the Phoenician canals, built for their new capital at Carthage in the 9th century bc [Allen, 1998, p105], or the huge channels, such as one that is 40m (120ft) wide, the remains of which can be seen to the north-west of Lake Poopo in South America [Allen, 1998, pp12-13]. In Peru there was one aqueduct over 450 miles long that carried water over sierras and rivers [Allen, 1998, p57]. At the site of the original Mexico City, Tenochtitlàn, they built a 10 mile dyke around the city perimeter to prevent flooding of the city and to avoid pollution by salt, bringing in fresh water by aqueduct from the mountains in two channels: one for consumption and one for cleaning. Gates in the dyke could relieve surplus water [Allen, 1998, p79].

The organization of water was an essential part of city infra-structure, whether for sacred or practical purposes. Temples nearly always had a close connection with water and a key role in the storage of water. This water was either in the form of an artificial or natural lake, or a sacred spring, and it was referred to as the 'watery abyss', the *abzu* or *apsu* [Leick, 2001, p3]. The 'sacred lake' was a useful reservoir in times of drought, in addition to providing water for ritual. Thus, we find a 'bath or artificial pool nearly 40ft long and waterproofed with bitumen', with broad flights of steps into it inside the citadel at Mohenjo-daro in the Indus valley [Hawkes, 1973, p273]; or stone-carved rectangular tanks and circular basins in the temple of Hathor at Serabit el Khadim; or a vast water cistern under the Hypogeum of Hal Saflieni on Malta which continued to be used long after the sacred site itself had been abandoned.

Archaeologist Brian Fagan describes the well-planned system in South America. When, for instance, the Maya built at El Mirador 'a maze of pyramids and plazas' on the summit with quarries below, the quarries captured the rainwater, while the plazas acted as 'catchment pavements to funnel water into them'. Then there were 'gravity canals that released water from the elevated central reservoir system into tanks and surrounding irrigation systems'. In effect, the Maya built their own 'micro-watersheds' to take full advantage of what rainfall there was [Fagan, 2004, pp233-4]. With effective water management, cities could sustain life in relatively harsh environments, such as at Machu Picchu high up in the mountains.

Ancient civilizations had drainage systems for taking away foul water from domestic buildings: in the Minoan town Akrotiri on Thera, north of Crete, there were clay pipes from lavatories into a public system [Fitton, 2002, p168, p171]; or in the cities of the Indus valley, the urban drainage was 'unparalleled in pre-classical times', in the view of some. In the Indus valley every house had a bathroom and privy, and there were earthenware pipes carrying waste from upper and ground floors into public drains below the centre of the streets [Hawkes, 1973, p276]. Efficient drainage and control of sewage has to be the mark of a civilized people. Without it large numbers of people cannot live happily in relatively confined spaces.

Large numbers of people are also likely to coexist more peace-fully if they share a common legal framework. One of the most fundamental laws for a civilized society is the law of contract. It is significant that both trade and justice share the symbol of a pair of weighing scales. Justice needs to be applied in a balanced manner, and equally trade relies on accurate weights and measures, as well as on the use of contractural agreements..

Trade

There have been amazing trade routes over the millennia. Perhaps the most famous was the Silk Route between the Middle East and the Far East, although the latter did not really get going until relatively late, having its heyday from around 1000 BC to about AD 1500. The Silk Route was typical of civilized trade in that it involved the exchange of goods along the route, together with networks of agents [Reid, 1994, p11].

Confirmation of the use of agents came in 1933 when French archaeologists excavated the site of Mari, modern day Tell-Hariri, at a midpoint on the Euphrates. A scale-model of the palace can now be seen in the Louvre in Paris. In the ruins of this large royal palace with as many as 300 rooms, dating from around 2000 BC [Phillips, 2002, p73], they discovered documents dating to the 18th century BC referring to goods from 'Kaptara', the ancient name for Crete, and to tin being for an agent, a man from Kaptara who was resident at Ugarit, a Phoenician city on the coast of Lebanon [Fitton, 2002, p99]. The tin, important for the manufacture of bronze, may well have come from as far away as Afghanistan, arriving at Mari before being sent on to the Phoenician trading ports on the Mediterranean east coast.

The Phoenician king Hiram was known for having opened up trade routes to the East and having agents in places like Babylonia, Uruk and Ur [Aubet, 1993, p102, p72]. As early as the fourth millennium BC we can see traces of long trade routes in Mesopotamia. Between about 3800 and 3200 BC, at least one city in the south of the country, Uruk, had links with northern Syria, southern Turkey and western Iran [Leick, 2001, p34]. Routes were established from Mesopotamia by sea along the coast eastward to the Indus valley, and west along the Arabian coast as far as Egypt. The Indus valley connection with the Arabic coast must go back a long way as there are eighth millennium Indus valley settlements which contain seashells from the Arabian coast [Mithen, 2004, pp408-410].

The main difference between 'civilized' trade and other forms of acquisition is thus one of mutual agreement. Trade is essentially a legally or morally contracted exchange of goods, rather than the chaotic seizing of booty as a consequence of warfare, banditry or piracy. Given the importance of trade to the

Phoenicians, it is not surprising to find that their Temple of Melquaart in Cadiz in southern Spain was used as a bank. Religious sanctuaries were an obvious place for transactions to take place in a foreign country. Here traders could check the 'quality of merchandise, exchange equivalencies and weights... under the protection of the god' [Aubet, 1993, pp234-235].

Food

Trade has been the stimulus behind the creation of a more civilized approach to food. Partly thanks to trade, the art of cooking developed, enabling the great *cuisines* of the world to appear. Salt, essential to food conservation, has long been traded over extended routes. Much later, international trade opened up the spice routes from Southeast Asia to Europe bringing back exotica we now take for granted – pepper, cloves and nutmeg – improving the taste and range of what we eat. The ancient Egyptians were not alone in importing food products from around the Mediterranean and from further south in Ethiopia.

Cities were also able to produce cooked food on a significant scale through the use of more organized methods than can be found among hunter-gatherer societies. The evidence of large-scale bakeries and breweries in the excavation of pyramid workers' dwellings in Egypt is only one such example [Tyldesley, 2000, p54]. But cities can survive only if they can rely on a sufficient supply of fresh food from the surrounding countryside, which means highly productive, organized agriculture.

Arable Farming

The agriculture which supports a city differs from any kind of hunter-gatherers' early farming in one simple respect: it requires the farmer to stay in one place rather than to follow the migration routes of the nomad. The farmer is thus responsible for the quality of his soil, fertilizing it and protecting it from erosion. If he invests effort and looks after it, then he will have good pasture for his beasts and better cereals with which to feed humans and animals. The other element that is essential to his success is water: too much and everything rots; too little and everything withers. By knowing how to use irrigation, the farmer can overcome such obstacles. The farmer thus has a close relationship with both the earth and the sky – especially the sky because the weather is so critical to success or failure.

Irrigation was essential to farming in Mesopotamia, where the rivers were not controllable, unlike the Nile in Egypt. From the fifth millennium BC onward there is evidence of terraces, canals and reservoirs for water management in Mesopotamia. Some of the canals were so large that they could take small sailing ships [Baigent, 1994, p24]. In Egypt, farmers could make use of the Nile and its regular annual inundation to water crops and provide fertility. Even so, there is evidence that an ancient dam existed near the modern Aswan dam in the south of the country; and it has been suggested that the lake of Moeris was created as a reservoir for the Nile, 450 miles in circumference, 350ft deep, complete with floodgates, locks and dams for irrigation [Donnelly, 1950, p185].

The Maya in Central America were typical of the model in their use of terracing on hillsides and in creating grids of raised field systems that resembled the swamp gardens of the Aztecs. The Maya began draining and canalizing swamps in the 1st century AD with the result that eight to ten million Maya lived in lowlands which, writer Brian Fagan points out, was 'a staggeringly high density for a tropical environment with low natural carrying capacity'. The danger in all these societies was that they were sometimes too successful, enabling too many people to live in what otherwise would have been marginal areas for survival. Fagan attributes the collapse of the Maya, for example, to having 'crossed a critical threshold of vulnerability' when drought came [Fagan, 2004, pp234-236].

Metaphysics

It was to the priestly caste, however, that the farmer looked for help with interpreting weather patterns and natural cycles. Not only were temples physically centered within cities, but priests and priestesses were involved in every aspect of civilization. It is almost impossible to reconstruct ancient attitudes toward gender. If the portrayal of deities is any indicator of a more general attitude, female attributes and roles were valued as much as male ones, although recognized as different. Certainly there were few exceptions to the ruler being a male – Queen Hatshepsut in Egypt being one of them.

Priests supervised the foundation of cities, the construction of great engineering works and monumental architecture. Their temples sanctioned the trading exchanges. They determined the agricultural calendar and predicted the weather (they were even rain-makers). They educated the young and kept written records. They healed the sick and helped the dying on their way. They had this total responsibility for one obvious reason: they were the ones with knowledge.

The importance of the seasonal calendar to the farming year was emphasized in that many of the ancient gods – Osiris, Melquaart, Dionysius and others – had a vegetative aspect to them, in the sense that the death of the god in the winter and his resurrection in the spring was related to the seasons. Priests officiated at important rituals which involved the pharaoh or king's re-enactment of god-like resurrection ceremonies, such as the Egyptian *Heb Sed* festival. The relationship between the priesthood and the political power base was critical to the ancient concept of civilization.

It could be said that knowledge gave priests even more power. Their status was enhanced if, for example, they were the only ones who could interpret omens for the king. The downside of this simplistic interpretation of power-hungry priests was that their status was useless to them if, because the king didn't like their omens, they were the ones who were put to death.

Above all, they had a deep understanding of the importance of the soul (the *psyche*) and its role in reincarnation and immortality. For them the relationship with the spirit world was in the form of a dynamic dialogue, best encapsulated in the concept of cosmic sympathy as expressed in the Hermetic dictum *as above, so below*, meaning that whatever happened among the gods was likely to happen on earth. This concept of cosmic sympathy was not, however, fatalistic and did not deny free will. There was a fundamental belief among the ancients that, through manipulation of the higher metaphysical plane, human beings were able to affect change in the lower physical world and interfere with the natural order for our own benefit (civilization is after all not a natural state for human beings) [Baigent & Leigh, 1997, p28]. In this sense fate was negotiable with the gods and relied on the specialist knowledge and skill of the priesthood for success.

In all respects the sacerdotal role was to provide a link between the metaphysical and the physical, from the abstract to the practical. Through ritual and ceremony, priests created conditions for the contemplation of the knowable and unknowable. The priesthood was responsible for interpreting the divine plan with regard to civilizing people and providing the fundamental harmony so that everyone could live in *Ma'at*. In other words, the priests and priestesses were essential to implementing and maintaining civilization. The temple held all the parts in balance, as it was through the medium of the temple that the rest of society had access to sources of knowledge.

Communication

The priestly involvement in the abstract leads us to the key aspect which represents communication in the form of education and travel. Through a careful and disciplined process of initiation and education, the priests passed on their knowledge to a select few – although this knowledge took many forms, both

Thoth & the art of writing

spiritual and mundane, based on more than just written records. Organized formal teaching and study are an essential part of civilization, and the Egyptian god Thoth was credited with one important invention that helped promote the civilization project: namely, writing. The ability to read and write continues to be one of the characteristics that supposedly distinguishes a 'civilized' society from a 'primitive' one.

In Egypt the idea of education was well established. Temples had specific areas called a 'House of Life' with departments dedicated to subjects ranging from medicine to astronomy, mathematics and law [Baigent, 1998, p192]. Particular temples in different parts of the country also specialized, such as the temple at Ombos which was particularly associated with healing. The information available to priests could thus be both detailed and specific.

Networks and the Ability to Navigate

Communication is not just about ideas but about travel as well. There is evidence that the ancients knew that the world was round, and that they were educated in the subjects of geodesy and geometry (the word 'geometry', after all, means 'Earth measurement') [Allen, 1998, p5]. One example of this knowledge is an ancient map known as the Piri Reis map. This map, possibly based on much older maps, was found in Istanbul in 1513 by Piri Reis, a Turkish sea captain. What is remarkable about it is that it shows a projection starting from Egypt which takes into account the curvature of the Earth – a map-making skill that was not rediscovered until hundreds of years later [Allen, 1998, p118]. It can also be argued that the Great Pyramid at Giza is a scale model of the dimensions of the northern hemisphere, its construction indicating a clear understanding of geometry [Allen, 1998, p118].

The civilizers applied the same principles of accurate measurement and straightness to the setting out of routes and other infrastructure as to other aspects of the model. Thus, the building of canals, irrigation channels or roads has resulted in extraordinary engineering and surveying feats which were precise to the point of almost ignoring physical obstacles. We tend to think of the Romans when we think of straight roads, but even Roman roads were sometimes built on top of existing straight Celtic ones. The idea is independent of the Romans.

Peruvian roads were old even by the time of the Incas and were well built. They were often 7-8m (20-25ft) wide – such as the one from Quito to Chile – and made with pulverized stone mixed with lime and bituminous cement. Their builders cut through rock and filled in ravines in order to build them, as well as making suspension bridges [Donnelly, 1950, p241]. But perhaps the most dramatic example of communication lines are the lines in South America called *ceques*.

The most famous *ceques* in Peru are those belonging to the Nasca peoples who lived on the south coast of Peru 2,350-1,400 years ago. The accuracy of the Nasca lines is most impressive; when the National Geographic Society funded a survey of them in 1968, they found that many of the lines were so straight that 'the deviation was less than four yards in a mile' [Morrison, 1978, pp44-46, p81]. These Nasca lines have survived undisturbed for thousands of years because they are in an inhospitable desert area of the Peruvian coast. The Spanish invaders destroyed others in the belief that they were pagan. The Nasca lines and the more famous drawings, however, were rediscovered by Dr Paul Kosok in June 1941, chancing upon them when he was studying ancient irrigation [Morrison, 1978, p29, p32]. The thirty or so drawings in the sand – which include images of a monkey, spider, humming bird, condor, flowers and hands – can only be seen from the air and vary in size from a few feet to 600ft across.

It was thought initially that the Nasca lines had some astrological significance. But astronomer Dr Gerald Hawkins, the Smithsonian astronomer in Boston who studied them in 1968, ran a computer check and could find 'no significant correlation between lines and positions of any stars at any conceivable date in antiquity'. This conclusion was not entirely surprising given that fog or haze from the cold Peruvian current would have obscured the view of the stars in the Nasca period anyway [Morrison, 1978, p35, p80]. If the lines had been used as calendrical devices, then it did not make sense that they ran from north to south rather than east to west, or began at centers and radiated out in spokes.

The *Omphalos*

Where these lines – of the Nasca and others – were the links between ancient sites, the centers of these sites themselves had great importance, usually symbolized by a stone called a 'navel stone' or *omphalos* in Greek. This concept can be found in both in the Middle East and South America. The original name of Tiahuanaco, 'Taypicala', meant 'stone in the middle' [Morrison, 1978, p144; Fagan, 2004, p238]. The meaning of the name of the Inca city Cuzco was 'navel of the world' [Hancock & Faiia, 1998, p274], and the Incas called their land Tahuantinsuyo, which translates as 'four quarters of the world', with its base centered on Cuzco [Allen, 1998, p87]. Lines radiate out for hundreds of kilometers from the navel of the world at Cuzco. According to writer Graham Hancock, one of these *ceques* centered on Cuzco passes through that city from near the ancient sites of Machu Picchu, Ollantaytambo, Sacsayhuaman and then extends 'without deviation' through Lake Titicaca to the city of

Tiahuanaco [Hancock & Faiia, 1998, pp294-296].

In Egypt the concept of the *omphalos* was expressed in the *benben* stone, the most famous being at Heliopolis. Greek stones – the most impressive example of which can be seen in the museum at Delphi – in particular often had a lattice pattern [Pennick, 1996, p121], and were 'thought to symbolize the latitudinal and longitudinal grid on the Earth' [Temple, 1999, p260, p280, p638, plates 14-18].

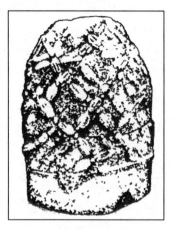

The *omphalos* at Delphi

Thoth in his form as Hermes, or Mercury with his winged sandals, has particularly come to symbolize travel. Piles of stones, referred to as 'herms' in his honor, are still frequently to be found along ancient routes. Thoth or Hermes is also connected with the final segment, that of medicine. As the Egyptian Thoth, he was closely linked with the art of healing, for example at the Temple of Thebes [Lindsay, 1970, p163]. His symbol of the caduceus continues to be widely used to represent medicine.

Thoth as Hermes or Mercury with caduceus

Medicine

With civilization came a sophisticated approach to medical treatment. The basic instruments they used remained almost unchanged for thousands of years. Medicine became more organized and systematized with developed repertories that recorded specific drugs. According to Clement of Alexandria (AD 150-200) there were 42 books attributed to Hermes referring to law, gods, priests and rituals of worship, including six books on medical matters [Merkel & Debus, 1988, p23]. Each of these six related to a particular area: anatomy, diseases, surgery, remedies and whole books dedicated to diseases of the eye and women's diseases. The *Ebers Papyrus*, the oldest known medical text, thought to date from 1500 BC, is reputed to be the book on remedies, book 40 in this collection. It is 68ft long and itemizes 811 prescriptions [Clark, 2000, p364]. Similarly, the *Edwin Smith Papyrus* contains 'significant evidence of a highly developed repertoire of treatment for the Egyptian surgeon' [Clark, 2000, p365].

To heal ourselves we attempt to rid ourselves of poisons, either real or imaginary, and thereby to purify our minds and our bodies. Priests, through their relationship with the divine, could heal the mind, the psyche, as well as the body with food and medicine. In ancient times there was no split, as there is today, between physical and mental illness – it was considered essential to involve the psyche in the healing of the body. There were special Egyptian priests called 'Senu' in Egyptian temples because a 'detailed knowledge of the spiritual origin of illness constituted much of the physician-priest's approach to treatment' [Clark, 2000, p364, p366].

The change from the diet of the hunter-gatherer to the more sophisticated and processed food of the city-dweller has, however, been a mixed blessing. There has been an increase in disease and decay as a result of making the change. It could be said that it was the domestication of animals that resulted in the transmission of infectious diseases due to the ability of bacteria to make the transition from animals to people (e.g. diphtheria from milk). Even today chickens are blamed for epidemics of human influenza. Food would appear to be both part of the problem and part of the solution.

The Greek god Aesclepius was believed to consider diet as essential to the treatment of illness. The Greek site at Epidauros maybe famous today for its amazing theatre but, in its time, the 4th century BC, it was better known as the place where people came to be healed with special dietary treatments by the god Aesclepius. While we might suffer less today from infectious diseases, we are still victims of ill-health due to our ambivalence about the appropriate amount of animal fat in our diets. We could learn much from our ancestors.

What should be clear from the above outline is that the theory of civilization was not invented by the Greeks and Romans in the first millennium BC. Those societies illustrate some of the basic elements of the archetype, and yet too many bits of it pre-exist in other ancient societies and on other continents for them to have invented it. Our Greek and Roman Indo-European ancestors knew of this archetype thousands of years before they became 'civilized', but they were not the original civilizers.

So, how did it come about? How did we make the significant shift from caveman to city dweller? It is usually taken for granted that cities grew from the exchange of agricultural produce. We interpret this as a progressive change because farming is, to us, obviously a more convenient and healthier way to live. Certainly, chronologically that is what would appear to have happened: first the Ice Age ended, then farming developed, then cities.

It may sound ridiculous to suggest, then, that with all our modern 21st century problems we should look again at what has happened since the last Ice Age. But if we want to know more about how this ancient archetype came about, as an alternative to the Greek and Roman ideas of civilization with which we are so familiar, we have to be prepared to go back a long way in time – beyond the Iron Age, beyond the Bronze Age, back to the end of the last Ice Age.

Chapter 3

Back to the Ice Age

The end of the last Ice Age was a moment of enormous change. 'Moment' is probably the wrong word, because it took several thousand years for floods to subside and for sea levels to stabilize as the world warmed up. But not only did the physical landscape alter dramatically as the ice retreated to reveal vast areas of land, but also something strange also began to happen to humans. About this time, 12,000-14,000 years ago, some of us began moving in a new direction. We began the all-important transition from hunter-gatherer to civilized man.

We point to evidence of the earliest farming, metalwork and pottery in the mountains of the Golden Crescent in the Middle East, as signs that humanity was moving forwards. It is easy to fall into the trap of thinking that this shift was inevitable and that Neolithic man was just waiting for the world to warm up in order to begin the farming experiment by throwing a few seeds around and penning some wild animals; or that some lucky individuals by sheer chance made a great leap forward, realizing that farming was a more successful way to live, thus starting a trend that began to change the world.

There are a number of problems with this scenario, not least because it fails to take account of what *actually* happened. We need to avoid making the cultural error of thinking that agriculture was 'progress' because it is what *we* know and *we* think of as an easier and better way of living than hunter-gathering. As a result, we omit to question certain fundamental aspects of farming such as the process of domestication or the skill levels involved. We fail to query the pattern of development of ancient farming as it took place so long ago. We accept without question that the trend to urban development must have arisen out of people's wish to trade their surpluses. Archaeologist Jacquetta Hawkes, writing about ancient civilization, realizes that such trading 'might have led to nothing more ambitious than the economy and cultural attainments of a market town' and not 'the higher flights of civilization' [Hawkes, 1973, p6].

By taking a closer look at this change so far back into prehistory, we can observe that, contrary to received wisdom, we did not make this shift *because* the Ice Age ended and the world got warmer. The shift was not inevitable. Furthermore, our understanding of how farming started does not stand up to critical analysis, neither does our usual idea about the arrival of cities.

There is no satisfactory conventional explanation for why civilization appears to start in unprepossessing mountainous regions like the Golden Crescent. Civilization was not necessarily the outcome of a human genius for experimentation and development, neither a natural curiosity nor a desire for self-improvement. It could even be argued that it was *imposed* upon a largely passive existing population – although who the original civilizers were can only be guessed at.

Couch Potatoes

The idea that humanity 'invented' farming, or anything else, underestimates our enormous capacity for making *no* change. If we take Europe as just one example, people who lived in Europe at the time of the Ice Age had a way of life that had not changed for thousands of years – the first Cro-Magnon settlers having arrived in Europe 45,000 years ago.

Those ancient people were happy enough. They had no particular reason to change. They had their basic toolkit for survival, their reindeer meat, their network of exchanges that stretched over vast areas. Their cave paintings show little evidence of aggression between them. Life was 'peaceful in the

Upper Palaeolithic'. There are 'no depictions of inter-human violence, and the few images interpreted as men holding weapons are confronting animals, not people' [Bahn & Vertut, 1997, p13]. For 15,000 years they even overlapped with the Neanderthals until the latter became extinct.

Agriculture does not appear in Britain for another *five thousand* years after the end of the Ice Age, not until 4700 BC. Is it not surprising that we did not invent farming, given that we in the West are supposed to be the ones who are so advanced, so highly evolved? We who in modern times seem to have been so good at inventing everything else? An interest in experimentation does not appear to be part of Europeans' ancient genetic makeup, at least not for the majority who might loosely be termed 'native' Europeans. It may come as a surprise to learn that over *eighty percent* of modern Europeans have maternal ancestors who were already living in Europe during the last Ice Age, in some cases for many thousands of years.

We did not evolve into modern couch potatoes: we are not fundamentally different from those early people who arrived as Cro-Magnons in Europe 45,000 years ago. Most people are still passive, happy to maintain a lifestyle they have always known. They like what they know and know what they like. When fashions change, the herd shifts together. To emphasize the point that we do not like too much change, it is well known that the descendants of Cheddar Gorge Man, who was alive 7,000 years ago, are still living in the Bristol area in southwest England. One has only to see how so many of us highly 'civilized' Westerners live in front of our TVs, vicariously living the artificial lives of soap operas while eating pre-prepared meals – in effect, allowing others to make decisions for us – to know that most of us are not prone to developing a big idea like farming. Indeed, Cro-Magnon lives on in us.

The reason for making such a claim is that Oxford professor of genetics Bryan Sykes has identified seven clusters of DNA in

Europe into which 'over 95percent of modern-day native Europeans fit' [Sykes, 2001, p244]. According to him, the 'single founder sequence at the root of each of the seven clusters was carried by just one woman in each case'. If this is true, it seems remarkable that so many people in Europe can be descendants of so few people, and that sufficient maternal lines have remained unbroken to allow genetic links to be traced so far back – mitochondrial DNA is inherited only through mothers.

But, according to Prof Sykes, by allowing for 'around one mutation in 20,000 years down a single maternal line' [Sykes, 2001, p197], it is possible to 'backtrack and work out how long ago two groups of people were once together as a single population'. On the basis that people with similar genes are likely to be related to each other, he has been able to observe in groups of people what he describes as 'genetic distance'; in other words, 'the bigger the genetic distance between them, the longer they had spent apart'. It therefore follows that 'genetic drift' occurs when there are 'bigger and bigger differences in the frequencies of genes as time passes' [Sykes, 2001, p64 & p63]. Using these techniques he has calculated that most Europeans are descended from just seven clusters, what he calls seven original 'clans' with seven foundation mothers, seven 'daughters of Eve'.

Of these seven clusters Prof Sykes has worked out that six were already present in Europe before the end of the last Ice Age. Indeed, he calculates that one clan, Clan U, arrived with the original Cro-Magnon settlers 45,000 years ago [Sykes, 2001, p263], while the bulk of modern Europeans, nearly 50 percent, are descended from a single clan, Clan H, that can trace its common maternal ancestor back 20,000 years; the rest, Clan V and Clan T, 17,000 years ago, and Clan K, 15,000 years [Sykes, 2001, p287]. Clan X was on the edge of Europe, north of the Caucasus, 25,000 years ago.

Significantly, only one clan, Clan J, whose foundation mother

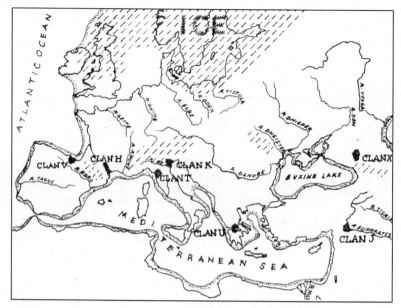

Map 1 The Ice Age: the locations of the clans; the full extent of the ice; the lower sea levels [reproduced with kind permission of Prof Sykes]

Prof Sykes calls Jasmin, had not arrived in Europe. This group stands out because it is the only one that hasn't genetically influenced the Basque country. The Basques are themselves intrinsically interesting given that modern Basques have the lowest incidence of group B blood of all Europeans, making them possibly the last descendants of the original Europeans [Sykes, 2001, p61]. Of interest is the linguistic legacy of the Basques' long ancestry – words for things like 'ceiling' literally translate as the 'roof of the cave'. Another reason for the Basques' genetic purity is that the Romans never conquered Basque country.

We, thus, have part of an answer. Farming seems to have come to Europe thanks to *one* group that was genetically different from the Ice Age hunter-gatherers. This last clan is clearly the key to the introduction of agriculture – not least because the clan originated about 10,000 years ago somewhere in Syria in part of the Golden Crescent where farming supposedly started. This group must have been the one with the farming skills – after all, Cro-Magnon man was more likely to drive an animal over a cliff than he was to start penning it.

Although only 15-20 percent of modern Europeans come from this clan, Prof Sykes realized when he traced it back that 'fully half of the Bedouin were in it' [Sykes, 2001, p183]. Prof Sykes has also traced this clan's route into Western Europe as it split into two. One lot, descended from Clan J, came up from the Balkans, across the Hungarian plain and river valleys of central Europe to the Baltic. They can be identified by their distinctive style of linear ware pottery (forever after identifying themselves as the LBK or Linear B and Keramik people). The other lot followed the Mediterranean coast to Spain, around the coast of Portugal and up the Atlantic coast to Western Britain. They are known for their impressed ware pottery. That early farmers followed both routes is confirmed in the archaeological record [Whittle, 1998, p139].

Lack of Experimentation

There is no archaeological evidence that the end of the Ice Age triggered a sudden interest in farming. We were not waiting for warmth. There had been previous warm gaps during the last Ice Age when there was no noticeable rise in creativity and inventiveness [van Doren Stern, 1969, p197; Sykes, 2001, p170]. There was only a faint sign of early farming *before* the end of the Ice Age in the Near East. When the Sea of Galilee dropped to an unusually low level in the autumn of 1989, a significant Early Kebaran site was found dating to c.17300 BC, which included many wild edible plant species and also basalt mortars and pestles and other bone and flint tools [Maisels, 1999, p92, p94].

If anything, by the end of the Paleolithic, Stone Age man elsewhere had started to decline. According to Clark and Piggott, the Azilian culture in southwestern France, dated to around 10500 BC, was 'almost in every respect inferior to that of the Late Magdalenian' – 'The flint workers developed no new ideas of importance' and 'The antler and bone industry was greatly diminished'. Furthermore, 'Though the Azilians almost without exception occupied the self-same caves, they apparently gave up blazoning their walls and roofs with representations of their prey' [van Doren Stern, 1969, p197]. Not much sign of progress then at the end of the Ice Age in Europe.

Furthermore, one of the peculiarities of prehistoric farming was its two distinct stages of development thousands of years apart, with no evidence of experimentation in between. The first stage, a sort of proto-farming that fits our image of early farming, coincided with the end of the Ice Age around the thirteenth millennium BC; and the second phase with the fifth millennium BC. It was this later stage when recognizable farming really happened.

The early phase first appeared in Egypt among the Palaeolithic communities of the Nile, c.12500 BC [Collins, 1998, p264], and then, c.9500 BC , moved to sites like Tell Abu Hureyra on the upper Euphrates in the Golden Crescent in northern Mesopotamia [Collins, 1998, p267]. The first 'fully morphologically domesticated cereals occur at Halula in the middle Euphrates Valley', c.7000 BC [Maisels, 1999, p106]. This stage may have involved the cultivation of wild cereals and the penning of semi-wild animals, but there were few other activities normally associated with farming, such as the milking of cows or ploughing. It was limited to the keeping of animals for their meat and hides.

While this first stage makes sense as an adaptation of hunter-gathering into farming, what we do not realize is that, had Neolithic Man wanted to try farming along the more familiar lines of today, he would have been unable to do so. Apart from evidence of domestic dogs in Natufian settlements dating from 12500 BC [Mithen, 2004, p34], prehistoric farm animals in general were not sufficiently modified to be of much use. The early sheep-goat type animal did not even have a woolly coat – early sheep having coats more like deer [Barber, 1999, p141]. Prehistoric sheep bones dating to 10500 BC found in a cave at Shanidar in eastern Iraq are from wild, not domestic sheep [van Doren Stern, 1969, p211].

This style of proto-farming remained unchanged, *unexperimented* with, for thousands of years. The Natufians, who were cultivating wild cereals from 12500 BC onwards, certainly didn't experiment. In the *three thousand* years that the Natufians were farming, they did not make the transition from wild to domestic cereals – even though we are led to believe that these same wild cereals were capable of being transformed into the domestic plants that humankind started to grow after 9500 BC.

Urbanisation – Çatal Hüyük

And if agriculture was supposed to be the stimulus for the building of cities, then it is odd that certain significant transformations of crops and animals, such as the domestication of the horse in the Ukraine, occurred thousands of miles away from such cities. It makes no sense that it is nearly *five thousand years* later, after the first appearance of agriculture in the Golden Crescent in the tenth millennium BC, that cities arrive further south in Mesopotamia. The time gap alone, of around 5,000 years, from early agriculture to cities, should arouse our suspicions. If cities were such an obvious outcome of farming, why did it take *so* long?

One particular archaeological site dating from the end of the eighth millennium BC in the Golden Crescent, Çatal Hüyük in Turkey's eastern Anatolia, clearly contradicts the idea that farming was the precursor to urbanization. Çatal Hüyük is fascinating. It contains some puzzling elements of the total civilization concept. To start with, it is clearly a town that covers a substantial area, complete with streets and houses, not mud huts. It is far from the usual Stone Age settlement, with 'highly polished obsidian mirrors, finely pierced beads, jewelry and textile work… including carpets' [Baigent, 1998, p118], as well as copper and lead casting. The obsidian mirrors had not a scratch; the beads had holes so fine that a modern needle could not pass through.

Given that Çatal Hüyük lies in the region from which farming began to spread, it is then something of a surprise to find that the diet of the inhabitants of Çatal Hüyük was resolutely Stone Age. It consisted of wild game, deer and aurochs. Çatal Hüyük was not alone in this respect. There is evidence of similar communities in the region, such as the 9,500 year old village of Tell Murebat near the Euphrates in Syria, which also consisted of close-built houses, yet still dependent on wild game [van Doren Stern, 1969, p207].

There are other anomalies at Çatal Hüyük such as the complete absence of pottery, in spite of the advanced style of jewelry and mirrors, sophisticated wooden and basketwork. Also the architecture is odd; it does not resemble the building work of later cities. It is still primitive and it doesn't develop into anything further – no later city grows out of its foundations.

The Disadvantages of Farming

Farming had to be *taught* rather than evolve through experimentation, for one critical reason: the experimenting hunter-gatherer and his family needed to avoid starvation through the winter. Anyone who thinks farming is easy should try it. Knowledge of farming is not innate. There is no smooth transition between hunter-gathering and farming. The skills of the farmer and the hunter-gatherer are completely different. To become a farmer is to become something else. As a way of life, farming entirely alters one's relationship with the natural environment.

Once Stone Age man accepted the culture-shock of being tied to one place in order to keep livestock – especially pigs – or to grow crops, he would have needed to learn a lot. He had to know how to observe the weather and the much longer seasonal cycles that apply to the planting and harvesting of crops, all of which is quite different from the rhythm of wild animal cycles, seasons and movements. For the first time in human history, Stone Age man would have needed access to a reasonably accurate calendar that could help him understand where he was in the farming year.

He would need to know about his soil: how to cultivate it, when to sow and when to harvest, and how to prepare for winter. He would need to know how to care for his animals: how to feed them properly to minimize the chances of their dying; which ones to slaughter: which ones to breed and keep feeding through the winter, and avoiding the temptation of eating them all. After all, with hunter-gathering, the herds look after themselves and do not require the involvement of people. The hunter-gatherer only needs to know how to kill animals, not how to keep them alive. Altogether Stone Age man would have to make a lot of effort to successfully implement his new lifestyle.

From our modern perspective, a hunter-gatherer lifestyle that depends on the haphazard chance of killing and eating something wild seems to us very risky. But in actual fact both ways of living are fraught with risk and neither has a guarantee of success. Wild herds can move too far away or crops can fail. One only has to consider the life of the modern-day subsistence farmer in sub-Saharan Africa to be aware of how fragile farming life can be.

Those who lead hunter-gatherer lifestyles do not necessarily see such lifestyles as hard. Studies of Stone Age peoples, particularly of those that have survived into modern times, show that in such societies there is a relationship between the hunter and the hunted. Wild animal tracks and habits are observed and learnt carefully. These people live in a fully symbiotic manner with their prey, with the flexibility to adapt and move on as necessary.

Changing away from hunter-gathering was also not in

people's best interest for other reasons. Hunter-gathering is in many respects a healthier lifestyle, and it was with the introduction of farming that we started to experience disease and decay. Lines of hypoplasias found on teeth – which are an indicator of food shortage because nutrition affects the development of healthy teeth – are less frequent in Natufian enamel dating from the middle of the thirteenth millennium BC than they are in later farmers [Mithen, 2004, p43].

There were also increases in dental disease in the late Neolithic – 'specifically *ante-mortem* tooth loss' – as a result of the growing reliance on cereals which are 'more abrasive and cariogenic' [Maisels, 1999, p92]. There are also fewer signs of degenerative diseases in the bones of hunter-gatherers that have been dug up. Conferences on palaeopathology that ask the question, 'Was the shift from hunting to farming more or less healthy?' frequently come to the conclusion that hunters' health was 'markedly better' [Rudgley, 1998, p8].

Unevolved Domestication

It would be interesting to know whether anyone has recently tried to naturally 'evolve' wild grasses into something approaching modern cereals. How many seasons would you have to wait before einkorn (a form of early wheat) began to taste vaguely of wheat and become useful as a food staple? Even more suspicious is the fact that what defines a cereal as domesticated is not so much the taste but the hallmark of civilization, namely *convenience* in harvesting and sowing.

In terms of cereal, the genetic change is in the germination of the seed and in the rachis, the hinge between the seed head and the stalk. In the wild plant the rachis is brittle and breaks easily in the wind, allowing the plant to spread its seeds as soon as it is ripe. A domestic version with a stronger rachis waits for the harvester to pick it. Likewise, a domestic seed waits to be sown before germinating. Daniel Zohary, a geneticist at the Hebrew

University in Jerusalem, who specializes in wild and domestic cereals, has established that this helpful difference between wild and domestic plants is the result of the mutation of a single gene [Mithen, 2004, pp23-24]. Those that have conducted mathematical modeling of einkorn have worked out that it would take 20 to 30 years to domesticate it if the einkorn were ripe when harvested, and much longer if not. One expert in this area, Gordon Hillman, has calculated that the rare genetic mutant, the seed head without a brittle rachis, has a probability of occurring only once or twice for every '2-4 *million* brittle individuals' and that it would then take 20 cycles of harvesting for these non-brittle seed heads to finally dominate the crop [Mithen, 2004, pp36-37]. Given the rarity of these seed heads, why would anyone bother to wait, especially with hungry families to feed?

Twenty to thirty years may seem a short time, relatively speaking, in the context of prehistory, but not in terms of a person's actual life. It would have been a short-lived experiment and the rest of the family would have returned to trapping long before it was finished. Not surprising then that the deliberate hybridizing of domestic emmer wheat with a wild grass, *aegilops squarrosa* from the Caspian Sea, to create bread wheat is described as a 'rapid transition' [Fagan, 2004, p94].

Curiously, this characteristic of domestic plants to wait for human intervention before being able to continue with the reproductive cycle applies as much to einkorn in eastern Turkey as it does to plants in South America, where domesticated Mexican beans, for example, similarly wait for the harvester. In the case of Mexican squash, domestication had the additional benefit of producing quite a different plant altogether – larger, thicker and bright orange, not green, and much easier to spot [Mithen, 2004, pp277-278].

Docility

The domestication of animals was as strange as the modification of cereals. Not only did animals change shape – as prehistorian Steven Mithen points out [Mithen, 2004, p34]: 'All animal species become reduced in size when the domesticated variants arise' – but they also became *conveniently* and usefully docile. Is evolution capable of producing the necessary change in the fundamental nature of an animal, even if a cow is still recognizably related to an auroch (considered to be the precursor to a cow)?

Anyone who claims that farm animals evolved out of tamed wild ones has clearly never worked with animals. Taming might conceivably work with a jungle fowl having its wings clipped and being bred into a chicken – although even a chicken can be vicious – but not with a cow, let alone a bull or a horse. Even domesticated modern versions of these larger animals are still capable of killing a person and demand enormous respect. They are too powerful and dangerous to be capable of being bred in captivity from wild and then turned into the sufficiently docile creatures necessary for farming.

If domestication was so easy, why has the zebra never been domesticated? Zebras must be no more than striped horses. Or, how come that, 'as genetic studies have confirmed, the Barbary Sheep never produced a domestic cousin'? [Mithen, 2004, p492]. Is it not strange, then, that today there persist breeds of horse or cat that have never been domesticated and *are not capable of being domesticated*? Even Julius Caesar knew that wild aurochs could not be tamed [Fagan, 2004, p156]. So how can one believe the nonsense that hunter-gatherers managed to tame aurochs because they 'culled more intemperate beasts and gained control over the herd' when they came into close contact with them during droughts? [Fagan, 2004, pp157-158].

If it really was possible to tame wild animals simply by penning them, over how many generations would it take for them to become no longer wild? Why would anyone wait to find out? Surely, if you breed a wild animal with more of the same species, the result is still wild? Are zoos not full of wild animals even though many have been bred in captivity? As people who are involved in breeding know, to achieve a change in outcome, one has to breed one thing with another that is different from itself. Charles Maisels acknowledges this point in the case of cereals: '*only* removal of seed grains from their zones of natural occurrence and their introduction to new habitats could have led to the formation of domestic variants, for otherwise the self-sowing characteristics of the brittle rachis would merely have caused wild stands to perpetuate themselves' [Maisels, 1999, p91].

Take the example of the onager, the precursor of the donkey. Timothy Potts, writing on behalf of the Oxford University Committee for Archaeology, explains that it had been thought that onagers were the 'domestic equids' of the Sumerians, but then he states that 'more recent investigations suggest that they cannot be fully domesticated'. So they were hunted for meat and hides and captured in order to be 'kept in a semi-wild state' for cross-breeding with domestic equids [Potts, 1994, p44] – which rather begs the question, *where did the domestic equids come from?*

There is of course the usual 'chance mutant gene' explanation. We are given the impression that, around the time of these early experiments Stone Age man was able to spot a genetic variation in the wild herds he followed and was capable of realizing that a particular animal was carrying a mutant gene that one day would make a 'useful cow'. But while we might know what would make a 'useful cow', how could Stone Age man know what the desired outcome was? How suspicious that the outcome was so convenient and so useful.

Secondary Products

This key change in the domestication of farm animals, trans-forming agriculture, did not happen until the fifth millennium BC, curiously, about the same time that cities appeared in

Mesopotamia. This next phase in farming is understandably termed by archaeologists the 'Secondary Products Revolution'.[1] Only now, more than *eight thousand* years later after the end of the Ice Age, all of a sudden people were able to take milk from a cow, wool from a sheep, ride a horse, use a light plough, plant vines and so on – that is, carry out a form of 'civilized' farming that has not changed in essence until the present day.

This idea of relatively fast change is supported by evidence found at Jericho. Jericho, as one of the oldest known archaeological sites, almost continuously inhabited for 9,000 years, is obviously of great interest. Here there is evidence in animal remains that the change from eating wild animals to domesticated versions of the same animals took place in the space of 100 years, which is too short a timescale for an evolutionary process to have taken place [Knight & Lomas, 2000, p94].

The implications of the modifications that happened in the fifth millennium BC were enormous. The ability to use wool, for instance, meant that a completely new and more versatile textile industry could develop. Although there had been some use of plant fibers before this time, such as flax to make clothes, the greater advantages of wool fibers now opened up the possibility of carpet-making and many different weaving techniques. The ability to produce felt from wool fibers transformed the lives of those who lived on the steppes in Central Asia because they now had a material that was waterproof and windproof as well as lightweight and thus easy to transport. They could use felt for many items, including tents as well as clothes and even food utensils. They also now had horses with which to implement their new lifestyle based on sheep-rearing.

Admittedly, it took time for this change to have a widespread effect. While the earliest evidence for domestication of the horse found at the site in the area of Sredny Stog in the Ukraine dates from 4000 BC [Barber, 1999, p35], the breakthrough in the exploitation of the deep steppe did not happen until about 3500

BC, and the first wheeled carts did not appear on the steppes north of the Black Sea and Caspian much before 3000 BC [Barber, 1999, p158].

One begins to wonder about the horse, though, after reading that the evidence for domestication is based on broken horse teeth, believed to be broken as a result of the use of a bit in the mouth [Jones, 2000, p210]. The wear on teeth could have been because of poor, rough diet, not bits – even the most junior member of a pony club knows that bits do not affect the wear of horse teeth. On that basis the Dereikva horse could have been a wild horse that was kept for meat or skin and not domesticated at all.

Given the ability to use horses for riding from the fifth millennium BC onward, what is surprising is that it did not translate immediately into a herding, nomadic lifestyle that would have been a logical step for hunter-gatherers. At Sredny Stog in the Ukraine, where the discovery of the ancient horse happened, 'the presence of domestic pigs… points to a settled regime' [Mallory, 1991, p199]. The same pattern occurred several thousands of years later in another region strongly associated today with big seasonal migrations: Ferghana, the most north-eastern farming region of Central Asia.

This is an area known today for the Kirghiz tribes who drive massive herds of cattle and sheep down through mountain passes into the Ferghana valley from Serirechiye and the Tashkent oasis. Here one would expect to find that herding animals would long predate a fixed farming culture. But, even here, the archaeological evidence suggests that, from the late second or early first millennium BC, first came the Chust non-nomadic farming and, second, the transhumance culture of the steppe, in spite of farming coming later to this region than elsewhere [Gorbunova, 1986, p12].

If farming really did arise as the result of experimentation, why did it take so long for the Secondary Products Revolution to occur? Surely there would be evidence of failed varieties along the way? Why did only the later stage of secondary products affect every aspect of farming, turning it into something that we would still recognize today? If it was simply chance, why did chance strike only twice? If farming was just an adaptation of hunter-gathering surely one would expect farming to start with sheep and cattle, because herding flocks on the steppes would fit in better with the nomadic lifestyle of Stone Age tribes?

The Point of Agriculture

So what is the point of farming? If trading agricultural surpluses was not the motivation for urbanization, how does one explain Çatal Hüyük and other such places? One should not forget that the time of Çatal Hüyük was relatively soon after the end of the Ice Age and the disappearance of the woolly mammoth. Life was still fairly primitive: we were not long out of the cave.

Yet the people of Çatal Hüyük had access to an anachronistic sophistication – such as the beads they had in their possession which were too fine for modern needles, implying a technical ability that does not re-occur until ancient Egypt *five thousand* years later. How does one make sense of Çatal Hüyük's apparently contradictory combination of sophistication and the primitive? If the people of Çatal Hüyük did not develop their skills of their own accord, then from whom and where did they acquire their knowledge?

How does one explain the arrival of farming in only nine places around the world, with specific crops and animals? How can one explain the comprehensive change of the secondary products revolution? Does this fundamental change happen purely by chance at about the same time that we start to see the early signs of civilization in Mesopotamia?

The domestication of only certain breeds of animal in specific areas and the earlier changes in wild cereals that gave rise to arable farming were suspiciously too convenient perhaps to be mere chance. Somebody knew how useful a domestic pig might be, or what an improvement a woolly sheep would be. Local domestication also had to be the case because, for example, the zebu humped cattle found in the Indus valley '…do not occur in the Middle East' [Maisels, 1999, p197]. While it is true that the genetic modification of wild into domestic horses took place away from centers of urban development, this transformation in the Ukraine was not an isolated event but a fascinating reflection of significant cultural activity taking place much further south of the Ukraine, beyond the Caucasus and in the direction of Mesopotamia.

Certainly one important motive for encouraging Stone Age man to change from hunter-gathering to farming was that organized agriculture can feed more people. As Jacquetta Hawkes points out; food surpluses meant that 'the way was open for considerable sections of a rapidly increasing population to give part, or even the whole, of their time to the practice of crafts and professions' [Hawkes, 1973, p6]. Techniques of organized agriculture, such as irrigation or terracing, also enabled people to survive in what were otherwise difficult environmental conditions. But it could be argued that food surpluses were necessary because cities needed them and, without cities, organized agriculture would not have happened on the scale that it did. Farming happened *because* of cities, not the other way round.

Mesopotamian Master of the animals

By 2800 BC, for instance, 80 percent of Sumerians lived in towns and cities [Fagan, 2004, p5]. For Sumerian cities to exist on any meaningful scale, not only did large numbers of people have to be released from agriculture to learn new skills but, perhaps more importantly, they had to relinquish control over *their own* food production. It seems unlikely that enough people in the Stone Age would have been willing to abandon providing for themselves, unless they were confident that they had another means of earning a living that would enable them to barter or exchange with those still in food production outside the city from the moment of their first arrival. As well as knowing which trade to choose, a new potter, cobbler or blacksmith would surely need to achieve a certain level of proficiency in his new trade *before* making the transition from self-provider to city-dweller. After all, one only has to witness the effort and difficulty we have today in retraining people in our post-industrial age.

Even on this limited basis, it is plausible to suggest that the introduction of cities was *deliberate* rather than the result of any natural evolutionary process, and that they were pre-planned from the outset, given the importance they attached to surveying and measurement – as supported by the archaeological evidence of cities in Mesopotamia, Egypt and elsewhere. The artificial nature of a city is one reason why cities could not *evolve* out of agriculture. They are essentially counter-self-sufficiency units: they force people to be interdependent. Because we in the West have lived through the Industrial Revolution we forget that cities were once the basis for all forms of production, whether leather, glass, metallurgy or textiles, in order to provide all the inter-locking specialist trades that go to make up a city.

Furthermore, if agriculture really was the stimulus for the creation of cities, why were the earliest examples of farming located in such agriculturally difficult places as mountainous regions? Why does civilization seems to start in the mountains? Archaeologist Jacquetta Hawkes is one of the few to draw attention to this curious situation in her comment that civilization 'was not inevitable, for on the one hand men have lived on well-watered and fertile land without creating civilization, and on the other they have created civilizations in apparently poor environments' [Hawkes, 1973, pp6-7].

It makes little sense, for instance, that the Golden Crescent in the Middle East should have been the so-called 'cradle of civilization', given how mountainous this whole region south of the Caucasus is. So, why was the harsh terrain of the Golden Crescent so attractive for thousands of years to a group of people who were sufficiently 'advanced' to leave behind evidence of agriculture, metalwork and pottery before descending to the Mesopotamian plain to build actual cities? What were they doing there in the mountains?

Chapter 4

Why Civilization Mysteriously Starts in Mountains

The Golden Crescent is not an obvious place for the 'cradle of civilization'. Even a cursory glance at a map reveals that it is almost entirely mountainous. 'Crescent' is a good description of the shape of this area of the Middle East as it sweeps in a wide arc through several mountain ranges: from the Taurus mountains in Anatolia in the Turkish west; through the mountains and foothills to the north and east of Iraq; past the two lakes, Lake Van and Lake Urmia; through Armenia, Azerbaijan and Kurdistan to Iran and the Zagros mountains in the southeast.

Likewise, in South America there are cities high up in the

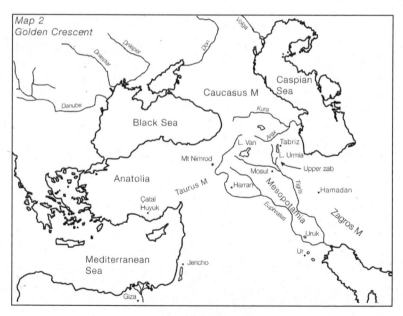

Map 2 The Golden Crescent

Altiplano or the sierras. Machu Picchu in Peru is an example of a city built high in the mountains. The civilizers appear to have chosen particularly unappealing sites for their early introduction of techniques – whether farming, metalwork or pottery.

All over the Golden Crescent there are no cities as such but instead the 'fingerprints' of the civilizers, with sites belonging to the Halaf appearing around the end of seventh millennium BC. Writer Charles Maisels describes this culture as 'remarkably (although not totally) homogenous'; it included pottery as well as other signs of civilization [Maisels, 1999, p141]. He comments on the fact that Halaf 'seems to have evolved... more or less simultaneously over much of its later range, negating any attempt to isolate its origins', having sites in southeastern Turkey as early as 6000 BC [Maisels, 1999, pp141-43]. The Halaf are thought to have traded obsidian, the black volcanic glass obtained from Nemrut Dagh (Mt Nimrod) to the north of Harran in the northern Euphrates and near the origin of Clan J in Syria.

The earliest cities in the Middle East were far to the south of the Golden Crescent, at the Sumerian sites of Uruk and Ur near the mouth of the Euphrates, and usually dated to the Ubaid culture some time in the fifth millennium BC. Between 5500 and 4500 BC the Ubaid culture took over the Halaf mountain sites and then spread southwards to nearby the city of Ur [Collins, 1998, pp312-13]. By 5000 BC there were several large communities in southern Mesopotamia with 2,500-4,000 people living in them, living on food produced by others. The Ubaid may therefore have been those early civilizers in Lower Mesopotamia.

Unlikely Civilization

Given that it only required the decision to use the Euphrates and Tigris for irrigation for farming to move to the Mesopotamian plain, it is strange to find so many traces and activities associated with the civilizers to the north in the mountains. Apart from hamlets with simple irrigation which can be found near Ur in the south as early as 6000 BC [Fagan, 2004, p5], the earliest archaeological sites are either in hostile mountain regions or in poor-quality foothills. Here there are traces of the earliest agriculture, the earliest ceramics, the earliest metalwork, even the earliest wine production – evidence of which archaeologists found at Hajji Firuz Tepe, south of Lake Urmia in Iranian Kurdistan. They discovered an earthenware pot with a dark residue dating from some time in the middle of the sixth millennium which, when analyzed, was found to have primitive wine residue – an ancient type of retsina! The wine grape, *vitis vinifera*, itself is thought to have originated in the Caucasus region near the Caspian Sea, north of the Zagros.[2]

Some writers have commented on the strangeness of these archaeological sites in the Golden Crescent which the records show have always been unlikely environments for farming. Diana Kirkbride, then director of the British School of Archaeology in Iraq, who in the 1970s excavated a particular site at Umm Dabaghiyah in north eastern Iraq in the Mosul-Sinjar region, dating to c.7500 BC, was struck by how 'singularly uninviting' it was still. She remarked upon the near-desert environment, a dry-steppe with no cultivated crops and almost no local water – the nearby swamp being salty due to deposits of gypsum. There was also a lack of trees for fuel for the same reason. The only resource locally in its favor was the availability of wild animals for hunting. Indeed, there were paintings of wild onager on the walls [Mithen, 2004, pp432-434].

Charles Maisels discusses another site that dates to c.6500-6000 BC to the east of Mosul, near Yarim Tepe on the Upper Zab, a tributary of the Tigris, where there is an interesting combination of cattle-breeding and hunter-gathering as well as wild and domesticated plants [Maisels, 1999, p126]. He comments that this site is in a 'zone of rocky limestone hillocks' that is '*not really suitable* for farming, yet there is plentiful evidence of it here' [Maisels, 1999, p125. Author's italics].

Also on the Upper Zab River is the Shanidar Cave at Zawi Chemi, overlooking the river. The Shanidar Cave is a very old site indeed. Archaeologists have found wild sheep bones from 9000 BC here, as well as an oval copper pendant one inch long taken from a human burial in the Shanidar Valley, with a carbon date of 8500 BC [van Doren Stern, 1969, p211, p298].

Here, in the Shanidar Cave, they have also found extensive evidence of a bird cult: large quantities of bird remains, including entire wings of predatory birds, carbon-dated to 8870 BC [Collins, 1998, p306]. Writer Andrew Collins links this site in eastern Iraq with Çatal Hüyük in central Anatolia and others in southern Turkey. At Çatal Hüyük there were whole rooms decorated with wall paintings of vulture-type birds [Collins, 1998, p307-09].

Shamanic Bird Cults

Similar evidence of a shamanic-style bird cult exists at two other sites in southern Turkey, one at Nevali Çori near Hilvan on the upper Euphrates, and the other further east at Cayönü. Sadly, Nevali Çori is now below the waters of the Atatürk Dam – built to produce hydroelectric power from the Euphrates – although much was salvaged and is in the museum at Urfa. Both of these sites are ancient, the one at Nevali Çori dating back to 8400 BC. It had basic agriculture, cereals and animals and was occupied at different times until c.5500 BC [Collins, 1998, p291]. What particularly struck Andrew Collins about these sites were their megaliths and their remarkable similarity to places in South America, especially Tiahuanaco in Bolivia.

Tiahuanaco, of course, was another marginal place for farming high up on the Altiplano and yet this city flourished for 600 years. Farmers succeeded by using techniques such as raised fields and lining them with clay in order to help them retain precious water. As Brian Fagan points out, 'such an elaborate city seems like a miracle in a cold, windswept environment with highly unpredictable rainfall, at an elevation where only hardy crops can grow' [Fagan, 2004, p240].

Another more gruesome parallel with South America is the evidence of human sacrifice found at these Turkish sites. At Cayönü, inside a building called the Skull Building, is an enormous one-ton stone block with the appearance of an offering table. Microscopic analysis of its surface revealed a 'high residue of blood' which came from aurochs, sheep and human beings [Collins, 1998, p298].

At Nevali Çori there were also signs of a shamanic bird cult. In all three phases of building, from 8400-7600 BC, there were statues of birds and bird men.

The Early Signs

It is, however, not far from Nevali Çori that there is evidence of the earliest cultivation of wild cereals and animal husbandry in the region, at sites like Tell Abu Hureyra on the upper Euphrates in northern Syria, where archaeologists found stone grinding tools and seeds dating back to around 9500 BC. During the three phases during which Abu Hureyra was occupied, between 9500-8000 BC, archaeologists have found basalt querns and grinding dishes and 'at least 157 edible seed species' [Maisels, 1999, pp100-01]. Again in Abu Hureyra, at a later date, c.7000 BC, they also found a large cockleshell with traces of green malachite used for eye makeup [Collins, 1998, p271].

It is at a later time, c.7000 BC, that we start to see the spread of early basic agriculture from Anatolia to other areas to the west and north. For example, when the earliest settlers arrived on Crete c.7000 BC they 'used types of bread wheat which are attested in Anatolia, but not, as far as we know, on the Greek mainland at this time', which is a fairly strong clue as to where they came from [Fitton, 2002, p39].

Not only did basic agriculture come from Anatolia but so did certain secondary techniques such as weaving wool. Although weaving using fibers like flax had existed in the Middle East since 7000 BC, wool could not be used because sheep had the wrong sort of coat, incapable of being spun until about 4000 BC. It was some time shortly before 3000 BC that we find the earliest evidence for the use of twill at Alishar in central Turkey, according to prehistoric textile expert Elizabeth Wayland Barber. The earliest use of felt also occurred in central Turkey around 2600 BC [Barber, 1999, p191].

Agriculture and its products, however, are not the only signs of civilization. I have already mentioned the existence of copper at Çatal Hüyük. It is in southern and eastern Turkey that there is other evidence of some of the earliest working of metal and the heating of ores to high temperatures to produce workable copper, from at least the end of the eighth millennium BC [Mohen & Elvere, 2000, p28].

In the Zagros Mountains to the east of Kermanshah, David Rohl has highlighted numerous finds of ancient pottery, particularly around Hamadan. In addition, at the Neolithic settlement of Jarmo, east of Kirkuk in Iraq, where evidence of farming has been carbon-dated to c.6750 BC [Baigent, 1994, p25], according to archaeologist James Mellaart they 'used pottery which appears without prototypes in the region' [Rohl, 1998, p136]. More pottery has been found at a site near Yarim Tepe, on the Upper Zab, which shows clear links with Jarmo in terms of design, dating to about 6000 BC [Maisels,1999, p129]. David Rohl makes the point that archaeology has 'shown that pottery was invented in the mountains of western Iran during the seventh millennium BC' and that 'the pattern of pottery shows a clear movement of culture from the highlands *down* into the lowlands' [Rohl, 1998, p136. Author's emphasis].

Eden

Indeed, so special was the Golden Crescent that there are even plausible grounds, according to archaeologist David Rohl, for believing that here was the location of the real Garden of Eden. My reason for referring to Eden is the curiosity of the location rather any intrinsic interest in biblical research. When the Jews in captivity in Babylon started to write what eventually became the Old Testament some time in the middle of the first millennium BC, using oral traditions and other records of the Babylonians, they had kept alive the knowledge of a place high up in the mountains that we know of as Eden for maybe as long as 2,500 years, so important was it to them and their ancestors.

The place that Rohl has identified lies next to Lake Urmia, a vast lake high up in the Zagros mountain range in northern Iran. He has worked out that the most likely location of the Garden is to the east of the lake where there is a fertile valley with fruit trees. At the eastern end of the valley is the modern Iranian city of Tabriz.

The Garden of Eden, synonymous with paradise, has long been a powerful symbol. It conjures up images of Adam and Eve, our perfect, untainted ancestors, drifting around in their innocent nakedness, enjoying the fruits of the garden, living in bliss. For biblical literalists this is it: this Garden of Eden is an archetype for a primal state of naïve happiness that we spend our earthly lives trying to get back to – God, after all has planted there every tree 'that is pleasant to the sight and good for food' [*Genesis* 2:9].

It has been downhill for humankind ever since Adam and Eve were banished from the garden for eating of the fruit of the Tree of the Knowledge of good and evil. If only the 'wicked' woman hadn't been so easily tempted by the serpent! This guilt-ridden view of our beginnings has been an essential part of Judeo-Christian teaching for thousands of years. On the one hand, we are provided with a glimpse of heaven on earth. On the other, it is taken away because we cannot resist temptation.

Although Darwin and others did much in the 19th century to undermine the literalists' belief in our origins, we are still left with our images of an earthly paradise, as well as prejudice regarding female weakness and temptation. But, regardless of whether Adam and Eve may have wandered around without any clothes on, or whether there were cherubim with flaming swords guarding it, there is no reason to think that the Bible is not describing a real place. It is rather disillusioning then to find out that, when Rohl actually visited the real life 'paradise' in the course of his research, he was not impressed and found it far from paradisiacal, for several reasons.

For a start, it doesn't help that the word 'paradise' has changed its meaning over the millennia. We might think of it as meaning 'Heaven on Earth' but the word 'paradise' is nothing more than a Persian word for 'walled garden'. It is not specific to the Garden of Eden as it applies to any garden with a wall. In this case it is arguably a description of the valley to the east of Lake Urmia because the entire valley is surrounded by steep-sided mountains, like a walled garden. The river that runs through it had an older name, Meidan Chay, which also has a meaning of 'walled' (as in Meidan-e Shah in Isfahan, meaning 'the walled garden of the king' [Rohl, 1998, p60]). Further confirmation that these are all just place names, toponyms with no particular religious significance other than much later attribution, is that the Miyandoab plain to the south east of Lake Urmia was referred to as 'edin' in Sumerian texts [Rohl, 1998, p102]. 'Edin' was just a name that meant 'plain'.

It is true that Rohl found a fertile valley full of fruit trees. The lake, however, he described as 'a melancholy place'. The current name of the river that runs through the valley, the Adji Chay, emptying into a vast salty marshland on the edge of the lake, means 'bitter waters'. Lake Urmia itself is salty and it reminded Rohl of another great, lifeless salt lake, the Dead Sea. Rohl writes that 'the desolation is oppressive' [Rohl, 1998, p104]. So much for paradise!

Why then did the *shape* of somewhere – a walled garden – give us our word 'paradise' when the reality of that place could not have been more different? What was the reason for spending so much time and effort in such difficult and unprepossessing terrain? What was so important to the ancients about inhospitable mountains both in the Middle East and South America? Oddly enough, *Genesis* itself gives one answer: *gold*.

The *Golden* Crescent

In mentioning the first river to flow out of Eden, the Pishon, *Genesis* states that this river flows around 'the whole land of Havilah, *where there is gold; and the gold of that land is good'*. David Rohl identifies the Pishon as the Kezel Uzun. According to him, the name *Uzun* means 'gold' or 'dark red'. He also cites the existence of a Sassanian gold mine from around the 3rd to the 7th centuries AD near the famous Zoroastrian fire temple of Takht-e Suleiman to the south of the alleged Garden of Eden. The river which flows from the extinct volcano of Takht-e Suleiman is called the Zarrineh Rud, which means 'Golden River'. As extensive later mining activity in the Golden Crescent demonstrates, the 'mineral wealth in the entire region is significant'; for example, recent mining for gold in the Ardabil region to the east of the Garden of Eden [Rohl, 1998, p56].

Mesopotamia to the south had a word for gold, KU.GI, but no gold itself [Potts, 1994, p164]. It all came from the mountains – either the Taurus Mountains to the north, or the Zagros Mountains to the east. When the king of Uruk made his demands in cuneiform tablets on the mountainous kingdom of Aratta in the third millennium BC for large amounts of gold and silver to be sent for the temples in the south at Eridu and Uruk, his requests may have been typical of an exchange that had been taking place for millennia beforehand [Rohl, 1998, p75, p130].

Apparently, there were also deposits of gold in the north of Iran along the Great Khorasan Road, and further south in western Iran in places like Isfahan. What is significant is that these deposits are often associated with sites where early pottery has been found. Another such place where gold was found is Hamadan and the area around it [Potts, 1994, p164]. Archaeologist David Rohl describes Hamadan, once the royal Persian summer capital of Medes, as being on an 'ancient highway leading to the exotic orient', including the sacred mountain of Behistun. This highway has mounds along the way of some of 'the most ancient pottery cultures in the Middle East' [Rohl, 1998, p87], and Rohl has traced a possible point of origin of this prehistoric pottery, dating to some time before the end of the seventh millennium BC, to a place called Tepe Guran, just south of this very old route [Rohl, 1998, p410]. Quite possibly gold deposits were the reason for the ancients being there in the first place – after all, pottery could be made from clay found anywhere.

Gold was no less important in Central and South America. The Spanish conquistadors were surprised to find so much gold used everywhere in temples and palaces. Tiahuanaco was just one example [Fagan, 2004, p238]. The Temple of the Sun at Cuzco was another. This temple was aligned to the midsummer solstice so that on that date the Sun shone directly into a 'trapezoidal niche', striking the seat used by the Inca monarch just as it rose. This niche was covered in emeralds, turquoise and gold plate [Morrison, 1978, p116]. But as we will discover in later chapters, there were uses of gold that were central to the whole civilization project which went beyond mere ornamentation.

Networks – Evidence of an Elite

On that basis, there are plausible explanations for pockets of more 'advanced' farming that the civilizers created in certain places, without requiring the building of actual cities. They needed to support operations being carried out in the mountains – not just the extraction of gold but also other metals and valuable materials like obsidian. Farming in Anatolia, as far back as 8300 BC, took place at sites close to sources of obsidian, and

Anatolian obsidian has been found along the eastern Mediterranean and the Persian Gulf [Fagan, 2004, p104]. For thousands of years, whatever useful resource was found in the mountains it was then moved elsewhere along established networks which lacked a centre. These early sites were therefore really no more than storage places, *emporia*, at key points along these networks.

Use as an *emporium* is a reasonable explanation for the unappealing site at Umm Dabaghiyah in northeast Iraq, mentioned above. It was far from being a typical Stone Age farming settlement. Although it existed slightly earlier than Çatal Hüyük, being before 7500 BC, it shared certain design features such as the appearance of being 'planned and built at once', and access to rooms was gained through the roof using external staircases and internal ladders. Several factors point to Umm Dabaghiyah's function as an *emporium*: above all, the importance given to storerooms built of better materials than the living quarters surrounding them. The storerooms formed the centerpiece of the settlement. They gave the place an organized feel, with lots of storerooms divided into smaller chambers, often with no doors.

In further support of the notion that these were storage places, there are indications that the 'bean counters' of civilization were busy from 8000 BC onward. Archaeologist Denise Schmandt-Besserat has examined over 10,000 pieces of a token system in the form of cones, spheres, disks, cylinders, animals, tools and so on, from all over ancient Mesopotamia, in use between 8000 and 4000 BC. She has worked out that this system was used for accounting and keeping records of animals. By 3700-3500 BC there was a refinement of the token system with a change to the use of clay envelopes that have the shapes of tokens pressed on outside [Rudgley, 1998, p48].

Another such emporium has been found at Tell Arpachiyah, also near Mosul in northeast Iraq – what Maisels refers to as an

'Anomalous Halafian building' [Maisels, 1999, p143]. The reason for this description is that many 'spindle-shaped dockets' and stamp-seals were found here, presumably used for trade, dating to the end of the Halaf period and the beginning of the Ubaid, between 4600 and 4300 BC. Charles Maisels describes the later Halafian culture as being a series of exchange relationships 'extending from the Zagros to the Mediterranean' with 'a site every 15-16 kilometers in the Mosul-Sinjar region'.

Circumstantial evidence suggests that it was the Halaf and the Ubaid who were closely involved in the implementation of civilization in terms of cities in Mesopotamia. Presumably what must have happened was that, some time after the beginning of the sixth millennium BC, the Ubaid decided to develop trading posts, the first cities, in southern Mesopotamia to trade with Bahrain and beyond to Oman.

Ubaid pottery has been found along the eastern shore of Arabia as far as Bahrain and Oman [Rohl, 1998, p410]. Oman was important to the Ubaid because it was rich in copper. Because of this trade, the southern Mesopotamian city of Ur became a significant port of entry for copper into Mesopotamia from quite early on [Allen, 1998, p117]. Eridu and Uruk were two other Mesopotamian early cities near the mouth of the Euphrates at which there are signs of the Ubaid culture. At Eridu there is evidence of 17 previous temples under the one dating from c.2000 BC, with the earliest dating possibly to 5000 BC or before [Baigent, 1994, p27; Collins, 1998, p357].

Not long after 5000 BC, Uruk grew to an impressive size, with life concentrated around the temple and the ziggurat [Fagan, 2004, pp133-4], and more than a thousand years later, it continued to have links with northern Syria, southern Turkey and western Iran [Leick, 2001, p34]. By 3600 BC Uruk was a great city with over 10,000 people [Jones, 2000, p211]. Uruk is an obvious example of a highly organized, civilized city, typical of the ancient model with its planned urban layout, its architecture, its pictographic script and its cylinder seals [Leick, 2001, p36, p45].

Yet in spite of the size and importance of Uruk, it was not a capital city. In fact the six Mesopotamian cities existing around the mouth of the Euphrates were all part of an alliance without a centre. Texts from that time refer to a mysterious central place called 'KI.EN.GI.' [Leick, 2001, p36, p78]. What is especially inter-esting about this name is that in Sumerian it roughly translates as 'land' (KI), 'lord' (EN), 'gold' (GI) – that is, possibly, 'the land where gold is lord' – thus emphasizing the significance of this metal to civilization, even though no gold existed in Mesopotamia. I will explain later why gold was of such impor-tance.

What we observe therefore in the Golden Crescent is not the evolution of civilization but the impact of early aspects of it. Çatal Hüyük does not develop further to become a prototype city. Human sacrifice of the type to be found at Cayönü and Shanidar are not part of the civilization model. Instead, what was part of it was the later domestication of farm animals, the transformation of deer-like animals into woolly sheep and the use of wool for weaving, the change of the cow into a docile creature providing humans with milk, the ability to ride a horse and keep herds, the use of metal, pottery and wheels, and so on.

The incongruities that clearly existed in the Golden Crescent only make sense if one accepts that the inhabitants of Çatal Hüyük and other places 'cherry-picked' certain artifacts and ideas from elsewhere: possession is not a sufficient ground to claim invention. In this respect they were no different from present day Stone Age tribes who sit in their traditional huts surrounded by the artifacts of their Stone Age way of life while watching television. To imply that the residents of Çatal Hüyük invented obsidian mirrors, tiny beads or copper casting is just as absurd as saying that inhabitants of Soweto in South Africa invented the television because they happen to watch it.

If neither evolution nor chance experimentation were the reason for civilization, then how did it occur? What is remarkable is that, wherever the *ur*-concept of civilization appeared, the ancients themselves attributed the changes to the influence of outsiders. Their legends, whether from the Middle East, Egypt or Central America, all refer to civilizers coming from elsewhere to teach them the art of civilization. Furthermore, there is compelling evidence that the civilizing skills we developed – the metalwork, the pottery, as well as farming – were *taught* skills.

Chapter 5

A Common Culture across Continents

Mesopotamia, Egypt and Central America shared more than pyramids. There are strong hints of a common culture, including legends of 'civilizers' and similar themes, in the Americas, the Middle East and Egypt. These themes, some of which we will meet again in later chapters, concern the connections between stars, the frequent references to 'watchers', the root word *Ur* with the meaning of 'foundation', 'oldest' or 'fundamental', and the widespread cultural importance of snakes.

Of course it could be argued that legends are no more than a social convention invented after the event, therefore valuable only as cultural devices. But however we might like to explain how civilization came about, the legends are consistent in referring to outsiders, gods, who brought civilization to the people – even if the names change and the gods are called Quetzalcoatl or Viracocha in South America, or Osiris and his wife Isis in Egypt and their companion Thoth, or Oannes in Mesopotamia. In particular, all of these civilizers shared one important characteristic: they all came in peace. They did not use force to introduce a new way of life.

Instead, they tried to persuade people to believe that civilization was a better way of living than hunter-gathering. As part of this message, they are reputed to have stopped the locals, even if only temporarily in some cases, from indulging in cannibalism as well as human and animal sacrifice, encouraging them instead to sacrifice only fruit and flowers [Hancock & Faiia, 1998, p5].

The description given in the *Hermetica* of Osiris and Isis could equally apply to any of the others:

Atum, the fabricator of the Cosmos, graced the Earth for a little time with our great father Osiris and the great goddess Isis, that they might give us the help we so much needed. They brought humanity divine religion and stopped the savagery of mutual slaughter. They established rites of worship, in correspondence to the sacred powers of the heavens. They consecrated temples and instituted sacrificial offerings to the gods that were their ancestors. They gave the gifts of food and shelter. Having learnt Atum's secret laws, they became lawgivers to humankind... and so filled the world with justice. They devised the initiation and training of the prophet-priests, so that through philosophy they might nurture men's souls and cure sickness of the body with the healing arts [Freke & Gandy, 1997, p88].

Isis

Osiris

These civilizers did not come alone. They had companions who were equally crucial to the civilization project. Both the Mesopotamian and Egyptian texts refer to groups of seven. At the temple of Edfu in Egypt there are texts that identify seven sages who were greatly venerated [Rohl, 1998, p336] as the Senior Ones, or members of the *Shebtiu*, and who, at the end of their time, 'sailed' away somewhere else to continue their creative task [Collins, 1998, p261].

The *Shebtiu* were the ancestors of the 'Followers of Horus' – Horus being the falcon god, son of Osiris and Isis. The 'Followers of Horus' (also called the *Shemsu-hor*) are usually shown in a line with either a bird or animal head and each holding a staff with a pennant on it (the Egyptian hieroglyph for deity, the *neter* or *neteru*, is a staff with a pennant). Interestingly, the Phoenicians referred to their Council of Elders as *sibutu* or *shibuti* until at least the middle of the second millennium BC [Aubet, 1993, p120].

Images of the Followers of Horus also appear beyond Egypt. An ancient Greek tablet from Phaistos shows what writer Rodney

Castleden refers to as an 'orderly but sinister procession of four daemons, each with its left arm hanging down and its right arm raised, holding a staff. One has a dog's head, another a boar's head, a third a bull's head and the fourth a bird's head' [Castleden, 1993, p143]. In his view these are 'sinister daemons' but it is just as likely that they are the *Shemsu-hor*. Even more noteworthy is the possible appearance of the *Shemsu-hor* in South America. The Gateway of the Sun at Tiwanaku on Lake Titicaca has three rows of 'winged functionaries with human or bird heads, each bearing its own staff of office' [Fagan, 2004, p240].

The *Shemsu-hor*, followers of Horus: detail from the 5,000 year old Egyptian Narmer palette

Star Worship

In terms of common themes, it is not surprising that there was a general fascination with stars, not least because ancient people were more exposed to the night sky than modern people. What is less expected, however, was the particular interest in the star system of Orion. One illustration of the importance of Orion is the curious fact that the three pyramids at Giza in Egypt and the three pyramids of Teotihuacàn in Mexico all appear to align with the three stars in Orion's Belt [Morton & Thomas, 2003, p161].

Another example of more general 'star worship' was a group of people in the Middle East for whom association with stars was so central to their way of life that their name comes from the Egyptian word for star, *sba*. These were the Sabians, identified with Harran on the northern Euphrates. Suffice it to say for now that their religion and perhaps some of their language came from Egypt. One Sabian word of interest was *ntr*. This word *ntr* can be spelt *neter* or in the plural *neteru*, which meant 'god' in Egyptian, and for the Sabians was translated as 'watcher' [Collins, 1998, p278].

This word 'watcher' often appears in the Bible. This is how the semi-divine beings who helped humanity are described in the Old Testament and other ancient Jewish documents. There are references to watchers in the *Book of Jubilees*, fragments of which exist in the Dead Sea Scrolls, in the *Book of Enoch* and in *Daniel*. In this context the Watchers are referred to as 'fallen angels' who have special knowledge that is passed to people [Knight & Lomas, 2000, pp152-3]. The *Book of Enoch* describes how these fallen angels married the daughters of men and 'taught them the arts of enchantments and various skills'.

The fallen angels, as a result of their involvement with their mortal wives, the 'daughters of men', also produced a race of giants known as *Nephilim* or *Rephaim*, described in the Bible as 'men of great stature'. Writer John Allegro considers the names *Nephilim* and *Rephaim* to mean 'fallen ones' (from heaven) [Allegro, 1981, p55]. Commentators Christopher Knight and Robert Lomas point out, however, that since *nephila* is an Aramaic name for the constellation of Orion, *Nephilim* must mean 'those who came from Orion' [Knight & Lomas, 2000, p155], a further link with that important star system.

The Serpent

Pyramids had associations not only with stars, but also with another persistent motif that appeared across the continents – that of the serpent. The founder god of the Aztecs, Quetzalcoatl or Kukulkan, the feathered serpent, was visualized in the Mayan design of their step pyramid at Chichén Itzà in Mexico – the light from the setting Sun on the solstice falling down the steps in such a way as to create the image of a serpent [Hancock & Faiia, 1998, pp26-7, p5].

Chichen Itza in Mexico

In the Middle East the snake had several guises with both negative and positive connotations. As a destructive serpent Typhon or as a symbol connected with healing. Toward the end of the first millennium BC there were groups, such as the Gnostic Ophites or the Naasenes, who were regarded as 'snake worshippers'. According to writer Jack Lindsay, 'the Ophites declared, "We venerate the Serpent because God has made it the cause of gnosis for mankind"' [Lindsay, 1970, p272].

In alchemy there has traditionally been the image of the snake biting its tail, referred to as the *ouroboros*, which can be found in the culture of several ancient peoples. Lindsay cites the view of one ancient commentator, Macrobius, that 'the world always goes rolling on itself in its globe-form... so the Phoenicians have represented it in their temples as a dragon curled in a circle and devouring its tail, to denote the way in which the world feeds on itself and returns on itself' [Lindsay, 1970, p267]. In ancient Egypt the same image appears in the form of Sito, the cosmic serpent, represented with many coils or with its tail in its mouth [Lindsay, 1970, p273].

The *ouroboros*

In Egypt there is no doubt that the snake was revered. The *Coffin Texts* state that 'The creative word was uttered by the coiled snake'. The *Book of the Dead* says that, at the end of time, 'The world will revert to its primary state, and Atum or Re will again become a serpent' [Lindsay, 1970, p273]. In the Egyptian creation myth found at Thebes there was also a serpent, a primordial snake called Kematef [Collins, 1998, p247], whose coils are depicted as the steps of the primeval mound (like the Pyramid of Kukulkan in South America). This Kematef had a close relationship with the primeval mound. He was considered to be a self-begetting snake who created the First Principle of Creation at the beginning – that is, the *'Bnnt* in Nun who fashioned the *bnnt* on the First Occasion' [Collins, 1998, p248].·I shall explore further the meaning of the *bnnt* in a later chapter.

The snake has long had associations with knowledge – not least the serpent's temptation of Eve in the Garden of Eden to eat from the Tree of Knowledge. When the Egyptian god Thoth eventually changed from his Egyptian form with an ibis head into his Graeco-Roman image of Hermes or Mercury, he is always depicted as holding a caduceus entwined with its double-snake emblem. Although this symbol has become associated with medicine and can be seen on chemists' shops in

The Kematef & the primeval mound as a pyramid

many countries, in fact it represents knowledge in general.

Ur

Hermes is the connection with the third theme, the important root word *ur*. For the Egyptians the word *ur* appears in their name for 'watchers' (the *urshu*) [Knight & Lomas, 2000, p155]. They had the same idea about watchers, believing them to be intermediaries who came between the gods and men. The Egyptians believed that these watchers came from a place called *Ta-Ur*, which means the 'oldest, far-off land', bringing civilization to Egypt during a very ancient epoch they called *Zep Tepi* – the 'first time' [Sitchin, quoted in Knight & Lomas, 2000, p156; Clark, 2000, p119].

Although the word *ur* for us is the name of Abraham's birthplace, Ur of the Chaldees (erroneously attributed to the famous ancient city that Sir Leonard Woolley excavated in the 1920s and 1930s in southern Iraq), it is no more than an epithet meaning 'foundation'. It is a word that crops up on both sides of the

Reed bundle boat in South America

Drawing of similar reed bundle boat on the Nile in Egypt

Atlantic. In South America, Lake Uru Uru, north of Lake Poopo, was the home of the ancient tribe of Uru Indians who lived 'amongst reed islands on the lagoon, building reed boats' [Allen, 1998, p78]. But it occurs most often in the Middle East.

Thus, one finds *ur* in Mesopotamian texts dating from about c.2600 BC in the word for copper (*urudu*) [Mohen & Elvere, 2000, p146], and in the original name of the Euphrates (*Urutu*) which meant 'copper river' [Allen, 1998, p116]. It appears in many place-names: *Uru-Unuki* was the city of Enoch [Rohl, 1998, p200]; the name 'Armenia' probably comes from *Ur-Mannai*; the Assyrians referred to the biblical Ararat as *Urartu*; Jerusalem was originally *Uru-Shalem* or the 'city of peace' [Rohl, 1998, p54].

Particularly intriguing, however, are the Greek derivations of two Egyptian terms which both contain *ur*. One is the Greek word 'pharaoh' which is thought to be a Greek interpretation of the Egyptian term *Per Ur*, which had the meaning in Egyptian of 'the exalted house' or 'house of foundation', referring to the sacred precinct at Nekheb close to Nekhen (or Hierakonpolis as it was called by the Greeks). It was here that the priestly caste known as the *Shemsu Hor* (or Followers of Horus) had a mission to initiate the ruler '…into the sacred mysteries of the throne so that he could guide Egyptian society to live in *Ma'at*' [Clark, 2000, p398]. The Greeks thus confused the function of a place with the role of a person for reasons that will become clearer later.

The other Greek word was their name for the Egyptian god Thoth, which has come down to us as 'Hermes'. Thoth was one of the most important in the Egyptian pantheon of deities who from the beginning was credited with many civilized inventions, from writing to medicine. Since *-mes* or *-mos* means 'son of' in Egyptian, then for the Greeks Thoth could have meant 'son of Ur' (*Ur-mes* or Hermes) [Lindsay, 1970, p321]. Much later on he was transformed into Trismegistus, the thrice-great Hermes, to whom the body of writing known as the *Hermetica* was attributed. As we will see again later, Thoth was not the only Egyptian deity to be found outside Egypt.

Pottery

Whatever the truth behind the legends bringing civilization, they point to the likelihood that something rather odd happened in our development as civilized people. No one can deny the skill-level involved in ancient crafts such as metallurgy or pottery, or in the building of something like the Great Pyramid at Giza. I am not suggesting that there were necessarily supernatural 'hand of God' events or visits from outer space – far from it. In my view there were elites who knew how to deliberately teach their skills to others, such as agriculture, pottery and metalwork. I shall explain later some ideas of how they obtained their knowledge.

It is the lack of evidence for experimentation in the use of these technologies that suggests that these were *learnt* techniques. Otherwise, how else can one explain the fact that the earliest examples are often the most impressive in terms of quality or complexity? Or the unvarying designs and consistency of production over many generations? Archaeologists express surprise at finding such levels of quality standard in the oldest artifacts because they assume progress through experimentation.

Pottery is one example of high standards at the beginning, which then declined. Presumably pottery developed because it was simply easier to make clay vessels rather than to work stone, and the technique was more transferable. The earliest pottery, which appears in the Zagros Mountains of Iran in the seventh millennium BC, is far from crude. It is delicate, often with striking patterns, in spite of being hand-made [Rohl, 1998, p137]. Similarly, writer Charles Maisels describes later Halaf pottery of the fifth millennium BC, which can be seen in the British Museum, as being 'aesthetically superb in both form and decoration' [Maisels, 1999, p137]. If anything, the quality of pottery production fell *after* the introduction of the potter's wheel.

Prehistorian Steven Mithen refers to pottery in the middle sixth millennium BC as being 'fine vessels' and by 4000 BC as coarse, 'wheel-thrown for mass production' [Mithen, 2004, pp412-3]. Archaeologist David Rohl also comments on the workmanship beginning to degenerate at about this time [Rohl, 1998, p34].

Expert commentators remark on the lack of variation in a product over a wide area. Standardization occurred in the Neolithic with regard to building work: archaeologists note the same kind of timber-framed long house outside Paris or in the Czech Republic, all with their doors facing east [Mithen, 2004, p180]. In Çatal Hüyük all the houses had the same floor plan: 'Even the doorways and bricks were standardized' [Fagan, 2004, p105]. Writers on Bronze Age Italy mention that the most striking fact about it was its cultural uniformity [Mallory, 1991, p92; Cornell, 1995, p31]. T J Cornell cites, as an example, its distinctive pottery with incised geometric design in sites 'hundreds of miles apart... with little or no visible variation of shapes or decorative motifs', and a similar homogeneity to be found in bronze tools and weapons [Cornell, 1995, pp31-2]. This uniformity can be seen in finds on display in the Natural History Museum in Vienna where the Bronze Age metalwork is so exactly replicated and well-made that it looks as if it had been produced in a factory.

Commentators find it odd that the technique – whether carved wall reliefs or wheel-thrown pottery – stays unchanged sometimes for thousands of years. Nowhere is quite as incredible as Egypt for the consistency and quality of artwork and hiero-glyphs, which remain the same clear symbols for thousands of years. In the Indus valley, pottery techniques are used today that still have the same mass production of fast, wheel-thrown pots and method of firing them since almost the beginning of the third millennium BC, nearly 5,000 years later. Between 500 to 1,000 unfired pots are stacked in layers between straw and given a roof of clay. The straw is set alight and burns for 24 hours. The pile is then left to cool for a week before the pots can be used [Maisels, 1999, p192 & p198].

These outcomes are explicable if one accepts that the early artisans were taught their trade and merely passed on their knowledge. Once people had been shown how to do something, they generally retained the same method unchanged for sometimes thousands of years, and kept that trade within the same family for generations.

Our assumptions about progress blind us to the fact that people also sometimes lost the knowledge of certain techniques. In Britain, after the Romans left, people forgot how to make a simple brick even though bricks had been used in building since the earliest times in Mesopotamia. It is only when in England the inhabitants had destroyed most of the oak forests to build warships during Tudor times in the 1500s, with so little wood left to build houses, that they went to Flanders to relearn the art of brick-making. The most spectacular example of skill-loss, however, applies to metallurgy, with the shift from bronze to iron.

Metalwork

Bronze, being an alloy, is a far more complex metal to produce than iron. Yet for thousands of years there were skilled smiths who knew in what proportions to combine metal to produce bronze – six parts copper and one part tin – but not how to smelt iron ore, even though iron is both easier to produce and in many respects more useful. Only limited amounts of iron, mostly in the form of meteoric iron, are to be found in use before the Iron Age.

As early as the end of the eighth millennium BC, people in eastern and southern Turkey had known how to heat ores to 1,000°C, producing copper oxides of malachite, azurite and cuprite [Mohen & Elvere, 2000, p28]. They knew how to achieve these temperatures before they had developed the art of cookery or pottery, which did not make an appearance until about 6300

BC. Yet the smelting of iron still did not happen until some time toward the end of the second millennium BC.

Metalwork was mainly for prestige and cult items, as well as for more practical goods such as buckets, helmets, pins and swords. In Mesopotamia bronze was considered to be a precious metal and noted alongside gold in a royal context at Ur in texts dating to 2600 BC. What is impressive is that people were prepared to cover vast distances to acquire metals for the making of bronze. Mesopotamia did not have any ore itself, but there is evidence of metal use there from the fifth millennium BC, acquired through Iran and the copper mines of the Arsarate region [Mohen & Elvere, 2000, p146].

Copper was fairly plentiful in the Middle East and Egypt. Indeed, Cyprus owes its name to the root term for copper (*cupreus* being Latin for 'copper-like') [Fitton, 2002, p24]. But tin was another matter. In the West, tin mines in Cornwall were exploited and supplied long trade routes for thousands of years, but these mines were not able to supply everywhere [Harding, 2000, p201]. Otherwise it is thought that tin came from east Uzbekistan or Afghanistan, and was then distributed onward from places like Mari on the Euphrates.

Apart from workshops found near temples, it is not known how metalworking was distributed among the wider population. There is no other evidence of permanent workshops as Bronze Age villages were too small to support their own smith, and yet there are occasionally hoards of broken and miscast metalwork. The explanation for this odd situation is that smiths were probably peripatetic, traveling from one place to another [Harding, 2000, p236], and thus producing to the same standard everywhere. Smiths were clearly keen to keep their secrets to themselves – a closed shop. The respect with which metalworkers and smiths were held should in itself alert us to the fact that metalworking was no ordinary skill that just 'evolved'. Even as far away as north-east Siberia, the Yakut regarded smiths as 'closely resembling shamans: they too can heal, give advice and foretell the future' [Baldick, 2000, p78]. In the Bible, Tubal-Cain, the founding father of metalwork, had an almost god-like status ('Cain' was a Hebrew word meaning 'smith' [Rohl, 1998, p200]).

Regardless of the truth or otherwise of legend – which is almost impossible to verify – there is nevertheless an unexpected source which confirms certain aspects of the *ur*-concept of civilization. This source is at least 5,000 years old. It was not directly involved in the establishment of civilization, but was on the periphery. What I am referring to is the analysis of the language of the proto-Indo-Europeans. They knew about civilization long before they became civilized, before the end of the fourth millennium BC – even before ancient Egypt became properly established.

There are no cities in any putative proto-Indo-European homeland. They were not the original civilizers. It would be more than 2,000 years before the Indo-Europeans began to live in cities. But at the time they came into contact with 'civilized' people – before 3000 BC – Mesopotamia was benefiting from the introduction of the 'total' Ur-concept. The Mesopotamians had domesticated animals, irrigation and knowledge of how to grow crops, the use of the potter's wheel, monumental architecture, written records, metalwork skills and gold. This was the dawning of the Bronze Age.

Chapter 6

The Proto-Indo-Europeans Confirm the Model

The origins of the Indo-Europeans are not well established. Nevertheless, analysis of proto-Indo-European language is fascinating. It not only confirms some of the original theory about civilization, but it is also through European linguistic ancestry that we have inherited our understanding of what civilization means. Certain relevant concepts have been embedded etymologically in European languages for thousands of years. Although archaeology is a useful way of backing up other data, potsherds cannot tell you what language someone spoke, or what they thought. From the study of palaeo-linguistics we can

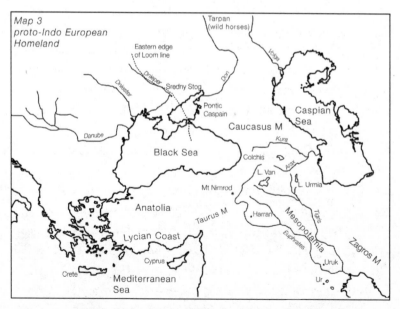

Map 3 Homeland of the Proto-Indo Europeans

reconstruct where and how people lived and when they lived there, as well as gaining some idea of what they knew. Extraordinarily enough, some of them absorbed disparate parts of the original concept at different times through their contact with the genuine civilizers.

Nineteenth century studies of the uncanny parallels between Indian Sanskrit and European languages – even though written with different alphabets – first alerted scholars to links between them. Given that so many words in Indo-European languages can be traced back to the same root, Indo-European tribes must have all once lived together, as *proto*-Indo-Europeans. Otherwise one has Monty Pythonesque images of ancient Celts traveling thousands of miles to let the Indians know that the latest word for 'x' is now 'y'. Hungarians, although physically quite close to these early Europeans, came from a different linguistic and tribal background, as indeed did the Basques. Both have remained as linguistic islands, providing a useful contrast and helping to prove that the links between the European languages are more than just coincidence.

The evidence suggests that the area where these different Indo-European tribes had a common homeland was the Pontic-Caspian, now part of the Ukraine, and that they last lived together over 5,000 years ago, until just before the end of the fourth millennium BC. It is the proto-Indo-European word for 'horse' (*ekwos*) that connects all the tribes both in time and place to the Pontic-Caspian. To find 'horse' remains in this region is not surprising given that, from the Dnieper River east to the Volga and possibly beyond, this is the natural range of the wild horse, the tarpan [Mallory, 1991, p162]. As already pointed out, the earliest evidence for the domestication of the horse occurred east of the Dnieper River – which flows south through the Ukraine into the Black Sea to the west of the Crimea [Jones, 2000, p210].

Another reason for taking the Dnieper River as the western boundary of their homeland comes from research into ancient weaving techniques. A line runs from the Baltic, down the Dnieper River through the middle of Turkey marking the eastern limit of the warp-weighted loom, which existed in Central Europe from 5500 BC, a basic weaving tool unknown to the Indo-Europeans before 3000 BC [Barber, 1999, p192]. Textile expert Elizabeth Wayland Barber claims that warp-weighted looms were unknown to early Indo-Europeans, who had their own names for different looms. So when they moved west of the Dnieper they borrowed from other languages the words relating to this new foreign technique [Barber, 1999, p191].

The proto-Indo-Europeans could not have lived very far to the east because their vocabulary does not reflect the desert or the steppe. Instead, they knew about plains, mountains and rivers, hot and cold weather including snow and ice, and three seasons of winter, spring and summer. They had names for deciduous trees such as birch, willow, ash, elm and oak, though it is less certain that they knew yew and pine. The wild animals and birds that they recognized, such as otter, beaver, wolf, bear, lynx, elk, red deer, hare, hedgehog, mouse, cuckoo, eagle, goose, crane and duck, definitely indicate a riverine or forest environment rather than desert or steppe. They were aware of bees and honey, which exist only in temperate climates in Europe or north Asia. They were not likely to have come from much further north because, even if 'sea' is not an obvious proto-Indo-European word, they knew about lakes. They also had words for boats and possibly oars [Mallory, 1991, pp114-117, p122].

Reconstructions of proto-Indo-European words connected with daily life show just how unsophisticated they were. They lived in simple houses where the main focus was the hearth. The house was possibly made of wattle and daub. At most they occupied fortified settlements [Mallory, 1991, p120]. They were not especially interested in any notions of social equality. They were firmly patriarchal: they had a term for 'widow' but not necessarily for 'widower'. The proto-Indo-European word for 'wedding' provides an insight into their social habits, as it means 'to lead home' – leading a wife to the home of the husband [Mallory, 1991, p123]. Some of them undoubtedly remained uncivilized in many respects for a very long time.

A Civilized Influence

Nevertheless, all of the Indo-European tribes were able to benefit from the Secondary Products Revolution in farming that happened in the fourth millennium BC. This was reflected in their many agricultural terms, such as 'grain', 'to sow', 'grinding', 'field', 'yoke', 'plough' and 'sickle'. It was then that ploughing became widespread through the Pontic Caspian, spreading to Europe, as far as Britain. They also knew about stock-breeding ('sheep' and 'cattle' can be reconstructed to proto-Indo-European, as can the secondary products 'butter' and 'cheese') and herding. Dogs, used for hunting or herding, attest to the 'earliest layer of vocabulary' [Mallory, 1991, p117, 119].

The sudden domestication of horses – sudden, because they had always existed in this area but had not been available before – was another beneficial change for the proto-Indo-Europeans. From the fifth millennium BC the use of the domestic horse seems to have spread westward from the Pontic-Caspian toward the Balkans, where there are to be found increasing numbers of horse remains. There was also the beginning of the use of wheeled vehicles in these areas – the proto-Indo-Europeans having several words for wheel, one for axle and for nave [Mallory, 1991; p121; Rankin, 1987, p30].

Metallurgy is another example of the influence of civilizers from elsewhere. The proto-Indo-Europeans did not know about iron (the Iron Age, after all, had not happened), and neither did they identify tin. Although they had no word for either of these two metals, they knew about bronze and they had a word for copper. What this suggests to J P Mallory is that the bronze was imported and 'the technique of its manufacture was unknown to the earliest Indo-Europeans'. Interestingly, the word for copper (*reudh) is linked to the earliest word for 'red' (the color of copper), and compares with the Sumerian word for copper, *urud* [Mallory, 1991, p121].

Perhaps most revealing is the Indo-Europeans' idea of a king. This concept must have been a straight lift from the model of civilization, giving a fascinating insight into the theoretical role of the ruler. Firstly, many Indo-European languages can trace their word for 'king' back to a common proto-Indo-European root, *reg, whether it is the Latin *rex*, Celtic *rix*, Indian *raj* or German *reich* [Mallory, 1991, p125]. Secondly, *reg is at the root of the word 'reach'. 'Reach' contains within it concepts of stretching out, that the king somehow delineates an important boundary by stretching out his right arm in a straight line as protector and ruler [Berresford Ellis, 1998, p49]. In Latin *rego* and in Greek *orego* both mean 'to stretch forth' [Rankin, 1987, p25]. Thirdly, *reg is the basis for accurate measurement ('regular') and the rule of law ('regulate') – all of which are key concepts in the civilization model. Maybe measurement was the reason for the stretching out of the king's arm in that the king set the standard length – as it were, from the tip of his nose to his fist – which is about a yard. English has preserved the confusion between the two concepts in its word 'ruler', which means both king and measuring stick.

What has confused linguistics is how a clearly civilized concept of kingship, based on the word *reg, with all its attendant meanings including the role of king as protector and 'measurer', could have existed among such warlike tribes. One explanation that J P Mallory has suggested is that maybe there was 'a leader who was more concerned with maintaining social and moral order than a secular sovereign exerting coercive power over his subjects or leading them into battle' [Mallory, 1991, p125]. In other words, two leaders! His reaction is typical of commentators who have been baffled by this curiosity. This unsatisfactory explanation is understandable only because commentators have not realized the connection between the Indo-Europeans and the true civilizers from elsewhere.

Religious Beliefs

The horse (*ekwos*) was not just a marker of the time and place of the origins of the Indo-Europeans but of Indo-European culture in general. Indeed, so culturally important did the horse become to the Indo-Europeans that it is the 'only animal name to figure prominently in the personal names of the earliest Indo-Europeans' [Mallory, 1991, p119]. They endowed horses with such special qualities that they sacrificed them – especially white ones. Horses were also buried with tribal leaders – no doubt to help them in their travels in the afterlife.

In Ferghana in Central Asia there is evidence of a Bronze Age Indo-European horse cult in the rock drawings at Aravan [Gorbunova, 1986, p195]. Ancient historian Herodotus describes how the Massagetes, related to the Saka tribes who lived in Ferghana, 'worship the Sun only of all the gods, and sacrifice horses to him; and this is the reason of this custom: they think it right to offer the swiftest of all animals to the swiftest of all the gods' [Herodotus, I, 216, quoted in Gorbunova, 1986, p195]. H D Rankin comments that the supreme day-god of Indo-European peoples was known in Irish as 'Eochu Ollathir Ruadrofessa', which translates as 'horse all-father red very wise one' and, given the color, sounds like a description of the Sun [Rankin, 1987, pp266-7].

More generally, many Indo-European languages had similar words for a superior god based on a concept of a 'sky father' or 'father from above', as well as a Sun god linked to horses [Mallory, 1991, p128]. The Indo-European words for gods were: Sanskrit = *devas*; Latin = *deus*; Lithuanian = *dievas*; Old Irish = *dia*. Indo-European terms for sky-fathers were: Sanskrit = *dyaus pita*; Greek = *zeu pater*; Latin = *Ju-piter*.

A solar cult also existed in Thrace in the late Bronze Age and, according to Sophocles, 'Helios [Greek for 'Sun'] was the most worshipped celestial body by the horse-loving Thracians' [Marazov, 1997, p43]. Certainly among the Greeks, Apollo, the Sun god, retained an elevated position in the pantheon of deities, as befits his name – which translates as 'the not-many', that is, 'the One'. The Romans under the emperor Constantine also did not abandon allegiance to their Sun god, *Sol Invictus*, in spite of conversion to Christianity in around the 4th century AD [Frend, 1984, p484]. We have continued with the idea of the day of the Sun, 'Sunday', being a special day.

The Celts had deities corresponding with Apollo, most notably Belenus or Belinos [Berresford Ellis, 1998, p54]. It is thought that Belenus may have the meaning of 'bright god' and was certainly linked to solar festivals such as the great feast that traditionally took place on 1st May, appropriately named 'Beltane' [Rankin, 1987, p266]. Reference to this god appears in the name 'Cunobelinos', according to Peter Ellis, 'One of the most famous Celtic kings in popular folk memory', in about AD 10. His name means 'hound of Belinos' and, according to Ellis, thence emerges into English literature as Shakespeare's Cymbeline. He was a powerful high king based at Colchester with influence over southern England, who died between AD 40 and 43 before the Roman invasion of Britain in AD 43 [Berresford Ellis, 1990, pp192-3 & p195].

A More Sophisticated Religion

It is not surprising that the Indo-Europeans shared a common reverence for natural phenomena, and especially for a host of deities reflecting the sky or the Sun. Their word for religion itself – related to 'creed' (Latin *credo* = 'I believe') – indicates an uncomplicated attitude to belief. 'Creed' can be reconstructed from two proto-Indo-European words, **kerd* and **dhe*, meaning to 'put in your heart'.

But what is surprising is the appearance in Indo-European

languages of deities from other cultures. J P Mallory makes the point that the Greek deities – Hermes, Aphrodite and others – are not obviously Indo-European Greek names, although he does not state where these names might have come from [Mallory, 1991, p67]. He does not realize, for instance, that 'Hermes' could derive from 'Ur-mes'. There are, however, more direct but equally unexpected references in several Indo-European languages to two major Egyptian deities, known to us Osiris and Thoth, as well as to the Egyptian word for 'god'.

Taking Osiris first, it must be remembered that 'Osiris' is his Greek name. His Egyptian name was *As-ar*, *Us-ar* or *Asar-uu* who, according to Egyptologist Wallis Budge, was a form of Osiris worshipped in lower Egypt [Temple, 1999, p131, p194]. It is curious, then, to find an echo of this name as far away as Bronze Age India. According to Jack Lindsay, there was a tribe of smiths called the 'Asurs' – significant in itself given the quasi-religious attitude to metallurgy in the Bronze Age – who 'seem to have lived in the North Punjab till expelled by Aryan invaders and driven to the mountains of Chota Nagpur'. He goes on to say that these smiths 'have been connected with the Asur and Asuras of the Vedic hymns...', which are among the oldest texts in ancient Indian history [Lindsay, 1970, p292]. H D Rankin also refers to *asura* in Sanskrit being 'used of a person with magical powers' [Rankin, 1987, p31].

Coming back further west, among the Iranians, *Asar* is transformed into the Persian mountain god *Ahura-mazda* who was revered by Zoroastrians for thousands of years and continues to be venerated in Iran and Iraq to this day [Rohl, 1998, p416; Rankin, 1987, p31]. On the western side of the Zagros Mountains, among the Assyrians – the name of whose very country was based on *Asar* – *Asar* becomes *Ashur* and many kings have names such as Ashur-uballit I (c.1365-1330 BC) [Leick, 2001, p207]. Both *Ashur* and *Ahura-mazda* share with their Egyptian counterparts the symbol of a winged sun disc: indeed, with reference to the god *Ashur*, 'on some Assyrian reliefs there appears a winged sun-disc with a bearded male wielding bow and arrows, that has been interpreted as a representation of *Ashur*, but this has never been confirmed by written identification' [Leick, 2001, p216].

The links between these Egyptian deities and the Celts are even more extraordinary. H D Rankin, in trying to find out more about Celtic beliefs, which only really exist in written form in the records of others such as the Romans, has discovered a 1st century AD poem entitled *Pharsalia* written by a Roman poet called Lucan. In this poem Lucan mentions 'a trio of Celtic gods' with the names Teutates, Esus and Taranis [Rankin, 1987, p264]. 'Taranis' is not possible to identify: he is unlikely to be 'Isis' (wife of Osiris) since one ancient commentator thinks that Taranis is Jupiter, god of thunder. The other two, however, are more obvious. 'Esus', described elsewhere by Rankin as a 'gaulish deity', could well be 'Osiris'. 'Esus' is credible as Osiris because of the possibility of 'Esus' having the same meaning as the Latin 'erus' which translates as 'master'. But there is much less doubt about 'Teutates'. 'Teutates' is unquestionably 'Thoth'.

Over the years there have been many spellings of 'Thoth', such as 'Tehuti', 'Thouth' [Merkel & Debus, 1988, p22] and 'Theutates' [Yates, 2001, p262-3]. What is striking is that the Celts were still calling Thoth by an archaic name in the 1st century AD when other Indo-Europeans, the Greeks and Romans, had already changed his name – the Greeks to Hermes and the Romans to Mercury. In further support of this connection, the ancient commentator on Lucan's poem is 'inclined to identify Teutates with Mercury' and, according to Rankin, Caesar reported that 'the Gauls particularly worship Mercury' [Rankin, 1987, p265, p260].

Another peculiarity is a word that in some of the Indo-European languages reconstructs as *netr*, a word that means 'god' in Egyptian (*ntr/neter*). J P Mallory states that it 'appears to be an innovation of the west Indo-European languages of Celtic,

Italic and Germanic' and cannot be ascribed to proto-Indo-European, thus implying that it is a later intrusion. Bizarrely enough, while in Egyptian it means 'god', yet in the Indo-European languages it has the clear meaning of 'snake' (Latin = *natrix*; Old Irish = *nathir*; Old English = *naeddre*, where it then becomes 'adder') [Mallory, 1991, p154]. Given the importance attached to the symbolism of the snake in the legends surrounding the civilizers, to find the coincidence of **netr* and 'snake' is quite fascinating. What is even more remarkable is that, for more than one tribe to have known these names and for them to have entered proto-Indo-European language, they must have known about these 'Egyptian' deities *before* the arrival of civilization in Egypt in the third millennium BC.

The Civilizers and the Caucasus

If the proto-Indo-Europeans absorbed certain notions of civilization into their language while still in their homeland, then there is only one plausible direction from which that influence might have come: from the south, below the Caucasus, from the region credited with starting civilization – the Golden Crescent. At the same time that momentous changes were taking place in farming among the proto-Indo-Europeans during the fourth millennium bc, the civilizers were active in the Golden Crescent. Their contact with the Indo-Europeans did not result in the building of cities to the north of either Mesopotamia or the Caucasus, but remained limited probably to an exchange of goods and ideas: the domestication of farm animals and secondary products, as well as concepts such as the role of a king and rudimentary religion, in return for resources like wild horses that could be tamed.

It is another 400 years before the horse makes a noticeable transition from the Pontic Caspian to much further south – by 3600 BC Uruk in southern Mesopotamia was known to have wheeled transport [Jones, 2000, p211]. Archaeologist Timothy

Potts makes the point that the horse is naturally an animal of highlands, and so not well suited to the extreme climate of the alluvial plains of Mesopotamia. Whether or not horses were used in the intervening period, between 4000 BC when they were domesticated in the Ukraine and 3600 BC when they appear in Mesopotamia, is difficult to say as they were not present in significant numbers in Mesopotamia until the late third millennium BC, after asses or donkeys [Potts, 1994, pp44-5].

For the civilizers to have influenced the proto-Indo-Europeans in the way that I have suggested, they must have come north of the Caucasus before 3000 BC because, by the end of the fourth millennium BC, there was a massive diaspora of Indo-European tribes south through the Caucasus. Some went west, in time becoming Celts and Germans, as well as Romanian and Bulgarian tribes. Some may have gone south and west, such as the Greeks. Others went south and east, becoming Iranians and Indians, and one group went even further east and became known as Tocharians.

The Caucasus themselves retained a powerful cultural signif-icance for one particular group of Indo-Europeans, the Greeks, long after they had migrated elsewhere. From 3000 BC onward, and possibly before, the Caucasus are known to have been the site of a flourishing early Bronze Age culture and possibly one of the places that copper – the basic ingredient of bronze – was first extracted, as the region is rich in malachite [Rohl, 1998, p64]. It was therefore of interest to the civilizers. Significantly, several Greek legends are located in this area. Robert Temple considers the legend of Prometheus, who is punished for giving heavenly fire to men by being chained to a mountain, then to be rescued by Hercules, to have originated in the Caucasus. He says it was in the Caucasus because of the importance there of mountains for geodetic and surveying reasons [Temple, 2000, pp156-7].

Then there is the Greek legend of Jason and the Argonauts. This tale is especially relevant because the ancient Greeks would appear to have anchored this archaic myth in a specific location at the foot of the Caucasus. Jason's quest to get back the Golden Fleece was from a place they must once have known, the land of Colchis on the border between Georgia and Turkey. Colchis is described as being in a broad estuary of the river Phasis with the Caucasus on the left and the plain of Ares on the right, where the fleece was to be found on a sacred oak with a snake watching over [Temple, 1999, p208, p213].

On the one hand, Jason and the Argonauts may well be firmly connected with the Caucasus; on the other hand, writer Julian Baldick has made a study of ancient religion in Central Asia. He has discovered remarkable parallels between this Greek myth and also that of Odysseus, with very old Altaic Turkic epics. The comparisons are convincing enough to suggest that these Greek legends have their roots much further to the east from a time that is impossible to date, given the use of oral traditions to pass these tales on. There is, nevertheless, a plausible timeframe with the tale of Odysseus, given that both Greek and Turkic tales refer to the reflex bow, which was in use by Central Asian Turkic nomads after about 3000 BC [Baldick, 2000, p164].

The legend of the Greek hero Jason compares with an Altaic epic called Maday Qara, recorded in 1964 at Gorno-Altaysk in southern Siberia.[3] The story of Odysseus reflects a Turkic folktale about a hero known in Uzbek verse as 'Alpamysh', recorded in the 1920s near Samarkand.[4] This influence on the Greeks of myths from the east could have occurred either while they were still in their Pontic-Caspian homeland, or later when they were *en route* between there and places later associated with them around the western coast of Turkey and in Greece itself.

It is more than likely that Greek-speaking tribes could well have spent time in transition south of the Caucasus. There are good reasons to believe that it was while in the land of Colchis that they encountered the Egyptians. While here, they may have heard the Odysseus story as it traveled west along the Silk Route, through northern Iran and into the southern Caucasus, modern-day Azerbaijan, along the Kuro-Araxes riverine routes before coming west to Colchis through Georgia.

The Celts

The impact of the civilizers on the Indo-European tribes before the end of the fourth millennium BC was not universal. The Celts are perhaps the most interesting, given that they were the ones who retained the most knowledge of the original model of civilization. For thousands of years other tribes such as the Germans remained 'wilder and more warlike'. In contrast to the Celts, the Germans had no Druids; they worshipped the Sun, the Moon and a god who could be Vulcanus; they did not have a developed agriculture; they did not consider robbery to be an offence; they did not have personal ownership of land but divided it up each year, even though the stronger could oust the weaker from their allocation [Rankin, 1987, p133].

Descriptions of the Scythians, one of the many Indo-European tribes who moved into Central Asia, from the middle of the first millennium BC, are equally unappealing. The Scythians became known as a nation of mounted archers who wore breastplates of bronze and formed a 'confused mounted horde' [Buttery, 1974, p44]. These horse archers do not present an attractive picture with their peaked caps, long unkempt hair and beards. They were particularly bloodthirsty, reputed to drink the blood of their enemies and hang enemies' scalps on their bridle reins. Ancient historian Herodotus refers to them as the people who 'scalp their enemies and make cups of their heads'. They never washed, using paste on their body instead – no doubt influenced by their time in Central Asia where it was customary to use smoke and fire for purification because, ironically, washing in water would then defile it [Baldick, 2000, p19, p89].

While the Greeks and Romans were much influenced by later civilizers with whom they came into contact, such as the

Phoenicians and Egyptians, the Celts of all the tribes may have had the longest-lasting memory of some of the original concepts. It is for that reason that we find traces of the original model of civilization in early Celtic society, which they passed on to the Romans. They contrast with the Scythians in being known for being clean, even giving their word for soap, *sapo*, to the Romans.

Even so, the Celts certainly still looked wild. They fought naked 'for religious reasons' in the belief that that would increase their unity with nature, and thus their power. Their nobles grew moustaches and shaved their cheeks, a practice frowned upon by the Greeks who took the view that 'moustaches were odd and probably unhygienic' – Spartans, for instance, were commanded to shave their moustaches [Rankin, 1987, p68; Berresford Ellis, 1998, p51].

The Celts believed in the importance of law and that 'the good of the community was the basis of law'. They elected their chieftains and officers. They were probably the least patriarchal of all the Indo-European tribes. Women had equal rights – they could inherit and own property. Women could even be elected to office and become leaders, as was the case with Boudicca. Celtic tribes also cared for their aged, their sick and their poor [Berresford Ellis, 1990, p16].

The impact of the original model of civilization is particularly apparent in the Celtic design of settlements, especially in the use of a cruciform basis for communities, some of which have now become large modern cities. Nigel Pennick, a Celtic specialist, has identified many modern cities that owe their foundation to the Celts – for example, London, Paris, Vienna, Bratislava, Aachen, Lyons, Belfast, Bregenz, Bologna and Milan. All of them are based on four principal roads that leave in cardinal directions [Pennick, 1996, p117].

According to Pennick, crossroads retained powerful meanings for the Celts. They represented places of transition in every sense. Crossroads were the point at which the cosmic axis intersects the underworld and upperworlds. The interaction between the spiritual and the material is symbolized in the Celtic cross, a sunwheel with strong geomantic connotations, similar to the Egyptian *ankh*. The cross was thus a 'universal pillar supporting the whole', the 'world axis' around which everything turned, mortal and immortal, referred to as *Yggdrasil* in Norse culture. Crossroads were sacred to the gods of travel, commerce and growth and so they were often marked with a 'Herm', an ithyphallic image of Mercury, psychopomp of the dead and sovereign deity of the crossways.

The Celtic cross kept alive the important idea of the spirit of place, the *anima loci*. It linked with the concept of the *omphalos*, the navel at the centre of the world, a concept shared with the Greeks. The Gaulish Druids viewed their meeting place at Chartres as their *omphalos* because it was 'center of their country'. This sacred aspect of crossroads is evident in the Celtic view of royal roads as sanctuaries, that cities on the crossing points of the ancient routes were small 'holy cities' in plan. For example, Watling Street, Fosse Way, Icknield Street and Ermine Street in England variously created crossroads at Cirencester, Dunstable and Royston. In many cases what we assume are Roman straight roads were actually built on top of pre-existing Celtic roads [Pennick, 1996, p135]. In this respect the Celts continued an ancient interest in sacred sites and the connections between them.

Using methods that compare with the study of genetics, we can work out roughly the order in which the Indo-European tribes separated. In other words, language groups that remained in close proximity tended to innovate more quickly. Greek and Latin, for instance, developed rapidly because the tribes speaking these variations stayed in relatively close contact; whereas the Celts, who migrated away first and moved furthest to the margins of the Indo-European world, often hung on to older style words and archaisms. What is extraordinary about the Celts is that they retained these concepts for thousands of years and were thus more 'advanced' than other tribes, in spite of being the furthest away.

According to J P Mallory, it is likely that 'the expansion of the Indo-European languages was broadly centrifugal from a more central homeland rather than lineal from one of the extremes'. He claims that the more central languages, from Greek in the west and Indic (including Armenian and Iranian) in the east, 'all appear to share late innovations'. Meanwhile archaisms are more likely on the periphery in Celtic, Italic, Phrygian, Anatolian and Tocharian. This pattern shows that the ancestors of peripheral language speakers had already dispersed before the innovations could have happened [Mallory, 1991, p155].

These connections between certain linguistic groups are demonstrated in the proto-Indo-European word for 'hundred'. One group, which included Latin and Greek, developed a hard-sounding 'c' (as in *centum*), while the other – Iranian and Indic languages as well as Celtic – evolved a soft sound (as in *satem*) [Rankin, 1987, p28]. These links explain why the Irish for hen, *cearc*, which has no equivalent in Western languages, is like *kark* in Ossetic, a language still to be found in the north Caucasus (the Ossetians were descendants of the Scythians).

The Celts kept other affinities with the Indo-Europeans who went to the southeast. One expert, Prof Myles Dillon, describes old Irish as being an 'extraordinary archaic and conservative tradition within the Indo-European one' [Berresford Ellis, 1998, p53], and close to the concepts of Sanskrit Vedas. He has found parallels in language, law, mythology and religious ideas between Celts and Hindus. In particular, the Celts shared the following characteristics with Thracians, Phrygians, and Scythians: ceremonial use of drink (Celts imported a lot of wine into Britain); raiding habits; 'wakes, funeral games, rejoicing at death as the inception of a better afterlife'; river and well worship; a three-headed rider-god, and pastoral emphasis [Rankin, 1987, p30].

Why the Diaspora?

For the Celts eventually to have arrived on damp, foggy islands, as far west as it is possible to go off the coast of Europe, must have meant that there was a serious reason to leave the original homeland. As I have suggested in an earlier chapter, people are generally reluctant to change. Certainly, the crisis that happened at the end of the fourth millennium BC was severe enough that none of the tribes seemed to have stayed behind. Before this crisis there was little reason to move.

One way or another, life was improving with the introduction of civilization. Beyond Mesopotamia there were civilizing changes. Some form of farming was being practiced in Britain and elsewhere in northern Europe. As we know from the evidence of the proto-Indo-Europeans north of the Caucasus, the Secondary Products Revolution in farming had taken off.

Even the climate was what one writer describes as a climatic optimum [Dunbavin, 1995, p87]. A transition had occurred in the climate at some point in the middle of the sixth millennium BC, from the Boreal period to the Atlantic period. Paul Dunbavin suggested that a subtle shift in the tilt of the world at an angle of 20 degrees meant that the temperate zones had been able to expand at the expense of tropical and polar regions. The tropics moved down to latitude of 20 degrees and below and the polar areas to beyond 70 degrees latitude [Dunbavin, 1995, p85].

All over the world the climate had become less harsh, more equable and temperate as the warmer and wetter Atlantic period began. Everywhere deciduous trees – oak, elm and lime – were able to spread their ranges much further north and south, taking over from pines. Deserts were also green – the Sahara was a green savannah, and had many shallow seas and lakes of which only Lake Chad now remains. The Gobi Desert was a forest. There were elephants and water buffalo in the Thar Desert region of northwest India and Pakistan [Dunbavin, 1995, pp90-91]. East Africa experienced 150-440mm rainfall annually [Fagan,

2004, p151]. There was also little difference between the seasons [Dunbavin, 1995, p154]. In fact, one could not get closer to 'paradise'. So, what could possibly go wrong?

The answer was *nature*. Just as we found ourselves powerless in the face of the Asian Tsunami in 2004 and other such events, and might so again in the near future, so were the ancients in the face of catastrophic natural events. In 3195 BC one such catastrophe struck the world with devastating impact.

Chapter 7

Disaster Strikes

No one is quite sure exactly what happened in 3195 BC. Whatever it was, it resulted in a massive social change that threatened civilization. This was the time of big migrations. What is interesting is to see how people responded and what action they took in order to re-establish civilization.

Although it could have been a volcano that blew up, some believe that it was a comet that hit the world in the eastern Mediterranean [Knight & Lomas, 2000, p289]. Since the Earth periodically passes through the Taurid meteor stream, material from this stream could have interfered with the Earth. Prof Schoch refers to a large asteroid called Olijato which is part of the meteor stream, and which 'from 3500 to 3000 BC ...had repeated close encounters with Earth', and therefore might have been the culprit [Schoch & McNally, 1999, p205]. Even a relatively small comet can have a dramatic effect. Whatever it was that affected the world, it had a terrible impact on everything, even geology, from people to plant life, which is one reason why one can be precise about the date through the use of tree ring analysis.

Irish bog oaks, dating back at least 5,000 years, show a marked change in growth rings at about 3195 BC. There is also a clear decrease in the pollen record at about this time, and in the lack of growth of bristlecones in the southern White Mountains in California, which have a chronology that goes back to 3435 BC. These high altitude trees are particularly vulnerable to drops in temperature below freezing and so they easily display evidence of frost rings [Baillie, 1995, p73].

Californian geologist Joe Briman refers to evidence of a slight return of ice in California and eastern Turkey between 3000 and 2000 BC and the re-advance of glaciers [Barber, 1999, p172]. There was a change in Neolithic Britain too after 3100 BC with the end of the appearance of a type of 'burial mound' known as long barrows (though whether or not they were built specifically for burial is a matter of debate); also there was a 250-year gap in agricultural development that resulted in forest regeneration in areas where previously there had been clearance.

Any major event which interferes with growing conditions is termed a 'dust-veil event' – normally attributed to volcanoes throwing up so much debris into the atmosphere that it blocks sunlight – and it is clearly recorded in tree rings, which are annual markers of the growth that has occurred in one season. Volcanoes blowing up and causing temporary havoc are not that unusual. We are all familiar with relatively recent eruptions such as at Vesuvius, Krakatoa and Mt St Vincent. These eruptions are typical 'dust-veil events': a few summers of bad harvests, and then normal life resumes. Destructive events, especially volcanoes, also reveal themselves in ice-core data from Greenland because the 'sulphur dioxide thrown into the stratosphere [from the volcano] arrives on the surface of the ice as sulphuric acid and is incorporated into the ice record'. The ice core data has a wider margin of error, however, than the tree ring record, and in this case it gives a date of 3150 BC (plus or minus 90 years).

Global Destabilization

But the difference about the event that happened over 5,000 years ago was that, according to dendrochronologist M G L Baillie, 'the identity of the volcano responsible for the 3195 BC narrowest ring-event remains unknown' [Baillie, 1995, pp91, 74, 78]. Furthermore, the whole world was destabilized by it, which would also suggest cometary impact. The effect of a comet would be powerful enough to create a sudden change in the Earth's magnetic field [Knight & Lomas, 2000, p72, p289] and shift the rotational axis of the pole, with the result that entire coastlines were displaced.

Writer Paul Dunbavin describes how an opposite effect occurred in each quarter-sphere of the globe. Only along a longitudinal 'line of neutrality' from Norway, Denmark, Italy and through the centre of Africa and back through Antarctica was there no change in sea level [Dunbavin, 1995, pp76-79]. In one quarter there were raised beaches, while, in the opposite, the land sank and forests were suddenly submerged in the sea. The suddenness of this event is evident in that these forests had no time to decay and were quickly covered in a layer of sand, because the increase in sea level was greater than the usual tidal range [Dunbavin, 1995, pp66, 68].

Not only did coastlines change, but rivers burst their banks and dramatically changed direction. Lakes also moved. Gwendolyn Leick refers to a major shifting of the Euphrates to its western branch around 3000 BC [Leick, 2001, p81]. David Rohl describes how Sir Leonard Woolley's famous excavations of the ancient city of Ur in Mesopotamia in the late 1920s and early 1930s came across a pre-flood level which showed 'every sign of being prosperous'. It was wiped out suddenly by fire and then experienced a great flood [Rohl, 1998, pp168-170]. It was quite possibly the time of the proverbial biblical Flood.

Admittedly, the flood of 3195 BC was not the only flood that ever occurred, either before or since. But what makes this event so unique is that its effects were so clearly global – from Irish bog oaks to Greenland ice cores. For example, is it just coincidence that the South American Mayan calendar gives a date of 3113 BC for their own time of crisis and the beginning date of the calendar [Rohl, 1998, p181]? Did the Indo-European belief, which the Celts held in common with the Hindus that 'in the time of primal chaos, divine waters from heaven had flooded downwards and soaked the Earth' relate to this period

[Berresford Ellis, 1998, p53]?

Did the Celts still retain an archaic memory of this appalling time when they met up with Alexander the Great in 334 BC on the banks of the Danube? Somewhat surprisingly, these brave warriors admitted to him that their one great fear was 'only that the sky will fall on our heads'. Livy, the Roman historian writing around the end of the 1st century BC, knew that the Celts remained frightened of 'such manifestations of natural power as thunderstorms, because they thought that they might presage the end of the universe'. Over a thousand years later this fear was still apparent in Irish law tracts in which the individual swore an oath to keep a bargain with the following words 'we will keep faith unless the sky falls and crushes us..." [Berresford Ellis, 1990, p76; Rankin, 1987, pp59-60].

The 19th century American Congressman, Ignatius Donnelly – famous for writing about Atlantis in the 1880s – is not alone in pointing out that there are flood legends from all around the world. While there is no proof that these legends all date from the same time, Donnelly makes clear that they share remarkably similar characteristics [Donnelly, 1950, p97]. This pattern of legend is consistent whether it applies to the biblical Noah, the Mesopotamian tale of Ziusudra, the Greeks' flood hero Deucalion, or the deluge myths common among many different Amerindian tribes.

In each case there is a founder of the race who is warned of a forthcoming disaster, while others disregard the warning. There is a vast conflagration which is put out by the flood that then follows. The founder hero escapes with his family. The hero's ship ascends the high mountain and he sends out birds in order to search for dry land. In each legend there is also an assumption that there are no other survivors. The floodwaters, however, all eventually subside and the survivors re-establish themselves on dry land.

Legacy of Disaster

Despite the legends' references to normal life returning after the Flood, there were two significant physical legacies of the catastrophe at the end of the fourth millennium BC which are still with us today. One was the widespread desertification of areas that once had been shallow seas or lakes – most notably the Sahara, the Thar and the Gobi. All of these places became deserts. In the case of the Sahara, only Lake Chad remained as an indicator of the much greater surface area of the water that once covered it. A more short-term, but equally serious, effect was an exaggerated wobble, or 'nutation', in the Earth's orbit. It is this wobble which, as we will see, has possibly left a legacy in a common decorative symbol of the time – the spiral.

The Earth is not a perfect sphere and does not rotate around an axis at 90 degrees to the horizontal – that is, to the plane of Earth's orbit around the Sun. The centrifugal force of the Earth's rotation itself causes it to bulge at the equator – there is a difference of just over 20 km in the distance around the equator as compared with the longitudinal distance around the poles. Since Earth is an 'oblate spheroid' or 'ellipsoid' with its rotational axis at an oblique angle, itself capable of changing, the poles of the Earth, when projected outward on the backdrop of the stars, naturally follow a circular path through space [Dunbavin, 1995, p26].

The angle at which the Earth rotates – the angle of obliquity, currently 23 degrees – means that the pole star gradually shifts in relation to the Earth's poles and the astral backdrop of the night sky zodiac also shifts slowly over time. This process, the precession of the equinoxes, takes a total of nearly 26,000 years for the Earth to cover the whole zodiac, and over 2,000 years to pass through each individual sign.[5]

But what could have happened more than 5,000 years ago meant that there was unexpected shift in the rotational axis, causing a more pronounced wobble than before. Dunbavin's view is that 'the imparting body which caused all this damage 5,000 years ago was probably a comet of about average size. The nature of the pole shift demands that it struck the Earth a near-tangential blow; most of the nucleus then bounced off into space again with its orbit radically altered' [Dunbavin, 1995, p162].

The Earth then became out of alignment with its new rotational axis and the axis that its shape naturally tends towards – like a spinning top that has been interfered with and starts to wobble – and the effect would have been massive earthquakes as the Earth's crust adjusted to the new rotational axis [Dunbavin, 1995, pp55-6]. For at least 50 years after the catastrophe there were repeated cycles during which the Earth's rotation followed an erratic spiraling path. Each time, it took seven years for the Earth to return to its previous position. The consequence was a variability in the length of the day and the year and thus difficulty in predicting seasonal change – in addition to changes in weather patterns.

During this period of change, according to Dunbavin, strange things would have happened in the sky, such as a daily change in the position of the North Pole. The height of the Sun and the Moon in Earth's sky would also have changed, with the effect that there would have been constant seasonal variations with either hotter summers and colder winters or cooler summers and milder winters. The increased wobble would also have produced significant tides that were longer than normal, and longer beaches as well [Dunbavin, 1995, p38].

The instability of the Earth's rotation had the effect of creating enormous uncertainty, even to the extent of possibly altering the length of the year by speeding it up. There is some evidence that the length of the calendrical year had originally been 360 days. Certainly, both the Mayans and the Egyptians had calendars of 12 months of 30 equal days. The number 360 was also symbolically important to people like the Mesopotamians, demonstrated in their seemingly arbitrary choice for the number of degrees in a circle – no mathematical imperative requires 360 degrees in a circle. The Egyptians deliberately inserted five extra days into the calendar, known as epagomenal days, to make 365. Interestingly, even in Shakespeare's time these days were referred to in England as 'Egyptian days'.

Spirals, Lozenges and Henges

Certainly, the ancients faced major challenges in re-establishing themselves. With chaos happening in the sky, one of the problems that populations faced, post-3195 BC, was the recalibration of calendars. Now that farming had become an important mainstay of life, the ability to predict seasonal change using calendars had acquired a greater importance than the mere observation of the migration patterns of deer, or other methods favored by hunter-gatherers.

In this context, two motifs occur repeatedly around this time on the Western fringes of Europe and on Malta, the lozenge and the spiral. They are usually dismissed as mere cultural decoration. There are, nevertheless, good reasons for believing that they had a more serious purpose. The spiral sometimes has 3½

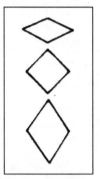

Right: Changes in the shape of the lozenge
depending on the line of latitude

The spiral as it appears on megaliths in Malta

turns and therefore could easily have been used to mark the progress of the seven-year wobble.

With regard to the lozenge, it probably relates to a technique involving a pair of wooden marker posts. The image of the Long Man of Wilmington carved into the chalk, holding a long staff in each hand, is a possible representation of this ancient technique. Using shadows from the two posts, referred to as the *asherah*, it is possible to establish the change in the height of the Sun through the year as it moves from equinox to solstice and back to equinox. Depending on the line of latitude, the pattern that the shadows make on the ground will vary from a square to a variation on a lozenge. People used this method for thousands of years to observe the movements of the Sun in order to work out, in conjunction with the phases of the Moon and one other easily observable heavenly body, usually Venus, exactly where in the seasonal calendar they were.

The importance of Venus was its use as the 'most accurate indicator of the time of year available in the solar system'. Every eight years, solar, lunar and sidereal (stellar) positions all coincide within few minutes, corresponding to five Venusian synodic years, and every 40 years all coincide within a few seconds. This allows three calendars to be realigned and prediction to be made of tides and eclipses, as well as being useful for feast days and agricultural purposes [Knight & Lomas, 2000, p266 & pp108-9].

One of the reasons why the locations of Stonehenge and Avebury in southern England were so important is because they were on a line of latitude, 51 degrees north, where the sunrise at the summer solstice aligns symmetrically with the sunset at the winter solstice [Knight & Lomas, 2000, pp178 & 333]. Certainly, there is evidence that post holes at Stonehenge for wooden marker posts long predate the stone monument. Stonehenge may well be one of the most famous of megalithic monuments today, but it was still in its earliest wooden phase at this stage, consisting of a circular earthwork with an internal bank and an external ditch [Bradley, 1998, p92]. Only later, between 2500 BC and 1600 BC, when there were constant reorganizations to the Stonehenge site, was the timber replaced with Bluestones brought from Wales.

Britain is not the only place where the *asherah* is to be found. For the Jews, *Asherah* was the Canaanite goddess who was the mother of the twins of Venus at dawn and dusk, as well as marker poles for gauging the movements of the Sun [Knight & Lomas, 2003, p184]. There are also frequent references in the Old Testament to the destruction of the *asherim* which, by the latter half of the second millennium BC had lost their practical meaning and acquired a quasi-religious status instead. Part of Josiah's reforms was that he destroyed the 'high places' where the *asherim* were sited and stopped people from burning incense on them [II *Kings* 23: 10-11].

In spite of the unpopularity of the *asherah* with Josiah, they remained an important cultural icon for a long period time, thousands of years after the crisis of the fourth millennium BC. The Phoenicians always placed two columns – of bronze or gold or precious stones – in the front of their great temples to represent their god Melquaart [Aubet, 1993, p35, p233]. Solomon was supposed to have had two columns at the front of his great temple, called respectively Jachin and Boaz.

Stone Calendars

Pairs of posts were not, however, the only means of recreating calendars after 3195 BC. It is from this date that there is significant activity in using megalithic stonework and the building of stone monuments that are in effect enormous timepieces on the western fringes of Europe at places like Callanish, the Ring of Brogar and Maes Howe in Scotland, at Anglesey in Wales and at Newgrange in Ireland, as well as the thousands of standing stones (40-50,000) along the western coast of France, for example at Carnac in Britanny [van Doren Stern, 1969, p247].

There was megalithic building before 3195 BC. Tunnel 'graves' in Western Europe are among the oldest structures made and some megalithic mounds in Brittany are more than 6,000 years old [Knight & Lomas, 2000, p235]. On the Mediterranean island of Malta there are megalithic temples predating the Western European monuments. The largest, at Hagar Qim, was built sometime between 3500 and 3000 BC. Even older evidence of megalith building comes from Egypt.[6]

Nevertheless, it is beyond coincidence that a series of stone buildings, all with one particular feature in common – long internal passages – began to appear on the western edge of Europe, as far north as the Orkneys, in increasing numbers around the date of 3195 BC. Inevitably modern archaeologists have interpreted their use, as always, as tombs, even though there are very rarely any bodies found in them [Knight & Lomas, 2000, p204]. It is also on these stone monuments that both the lozenge and the spiral motif can repeatedly be found.

What is far more likely, given their alignments, is that these passages were deliberately designed to receive light right at the back of the passages at the time of the solstices or equinoxes through an aperture at the entrance, or a roof box. By designing long narrow passages anything up to 100ft long, the ancients had probably created a 'collimator', a device that forces light rays to become parallel. Bryn Celli Ddu in Anglesey is one such 'passage tomb' with a roof slot where they could use observations of Venus to correct any solar or lunar drift, and where the light of the summer solstice illuminates a 'roughly carved spiral' on one particular stone slab [Knight & Lomas, 2000, p330, p259, p263].

Bryn Celli Ddu, Newgrange and the other sites were therefore quite specific to the need of that time, further emphasizing the seriousness of the crisis. Dunbavin is of the opinion that the increasing 'transition from passage observatories to outdoor alignments', at henges like Stonehenge, could have had much to do with 'the decreasing amplitude of the phenomenon [the seven-year wobble] over the centuries' [Dunbavin, 1995, p252].

Skara Brae – An Ancient Outpost?

There are several reasons for believing the builders of these strange stone monuments which, in some cases like Skara Brae in the Orkney Isles, include an entire village built of stone, complete with internal stone beds and dressers, knew about the ancient model of civilization. The people who had the skills to build them were known as Grooved Ware People after their distinctive pottery which often had lozenge and spiral motifs on it, unlike the round-headed Beaker Ware People who followed them. Their standard of skill level was extremely high. The masonry at Maes Howe, near Skara Brae, consists of enormous 5m slabs weighing three tons and precisely chiseled, accurately fitted and plumbed to the vertical. At Skara Brae there is also a complete underground sewage system [Knight & Lomas, 2000, p169, p187, p208, p203].

Skara Brae has other unexpected features. There is unusual evidence which would confirm its role as an observatory outpost – all the means for living on the island were imported. At the time Skara Brae was occupied there was no wood for burning on the island as Orkney was then mostly open grassland with few trees. The peat that now exists on the island was laid down 1,000 years *after* Skara Brae was abandoned in 2655 BC. Yet, they had sufficient fuel for the heat from the fires to crack the rock in the fireplaces and to fire up their pottery. Examination of Skara Brae middens – rubbish heaps – dating to this time reveal that the inhabitants ate sheep, cattle, pork, fish and oysters but no wild game, unlike other Orkney middens which have 50 percent game. They had no obvious facilities for keeping stock and, even stranger, bone-remains consist more of carcasses than skulls – all of which suggests imported pre-butchered carcasses [Knight & Lomas, 2000, pp186-7].

Clearly, these were not primitive people. Their choice of these sites on the western periphery was not arbitrary. Specific lines of latitude mattered to them in their astronomical observations for reasons no different to modern-day astronomers who choose to be in Yorkshire, Chile or Hawaii.

Celts go East

But while the clever Grove Ware People were recalibrating calendars in the British Isles, what was happening to other people? How did they react? Understandably, the first response of many people in 3195 BC was to move to somewhere safer. This became a time of big migrations, when proto-Indo-European tribes began to split and find their way to the eventual homelands with which we associate them. J P Mallory makes the point that the pottery style known as Corded Ware has a time horizon of 3200-2300 BC, with most dates in the third millennium BC, and is 'thoroughly congruent with the more immediate origins of the later Celts, Germans, Balts and Slavs, and possibly the Italic-speaking peoples' [Mallory, 1991, p246, p182].

Of all the Indo-European tribes who began to move at the end of the fourth millennium BC, it was the Celts who traveled the furthest – in both directions. Thus we find Celts as far west as it is possible to go, in Ireland, and almost equally as far east, in the Tarim Basin on the western edge of China in what is now the Takla Makan desert. The Celts who went east seemed to have followed what became the northern route of the great Silk Road, along the Kopet Dagh mountain range separating Iran and Turkestan, then up through the oases of Merv, Samarqand and Tashkent, finally ending up in the Tarim Basin, where they arrived around 2000 BC. The Tarim Basin, although a fearsome desert now, still had forests and swampy reedbeds left over from a time when it had been an inland sea at the time of the last Ice Age, and from the later temporary readvancement of glaciers, possibly as a result of the cometary impact [Barber, 1999, p173].

How we know that Celts went this far is that 4,000 year old bodies have been found, mummified in the dry, salty desert. These bodies are quite unlike those of any indigenous tribes from the Mongolian steppes to the north, or any Chinese groups to the east. They typically have Caucasoid features and are tall and fair or, in the case of the mummy found at Loulan, they have auburn hair.

DNA tests on at least one of the ancient bodies found at Qizilchoqa, dating from 1200-1000 BC, have shown links with the local race of Uighurs, who have distinctly Caucasian features, as well as with people from Central Europe [Barber, 1999, p193]. Textile expert Elizabeth Wayland Barber, who went to China in 1995 to examine these mummies, commented on the number of people living today in Xinjiang (the Chinese region which includes the Tarim Basin) with 'blue eyes and light brown or reddish hair' [Barber, 1999, p72].

Geneticist Steve Jones confirms that this genetic distinction is likely to have come from the west and not the east, pointing out that China has few variants in hemoglobin except along the route of the Silk Road. At the western end in China 1 in 200 people has abnormal hemoglobin, while at the eastern end 1 in 1,000 people has it [Jones, 2000, p213] – although this variation could be attributed to trade activity from the west over the millennia, as much as anything.

Tocharian

Another connection between these eastern Caucasians and the Celts are the remains of a language found in that area called Tocharian. Although written in Sanskrit, archaic documents found at several oases in the Tarim Basin, dating from the 6th-9th centuries AD,[7] reveal that Tocharian had much more in common with the Celtic, Italic and Germanic groups of languages than with Indo-Iranian [Barber, 1999, p116]. This suggests that these people had traveled east independently of other Indo-European tribes such as the Iranians. Although Tocharian seems to have died out after the 9th century AD, it left its legacy in toponyms for the Heavenly Mountains range to the north of the Tarim Basin ('Klyom' or 'Klyomo' being Tocharian for 'heavenly') – still known as the Qiliam and Kunlan mountains [Barber, 1999, p123]. (The Tocharian word for 'sky' has the same root as the Latin *caelum*, from which we get the word 'celestial'.)

What is slightly odd about these Tocharians is that they left their proto-Indo-European base without horses. Horses did not arrive in the Tarim Basin for another 1,000 years. One of the later mummies, the 2m tall Cherchen Man, who had light brown hair and a beard and who dates from about 1000 BC, is buried with a leather saddle and the head and front hoof of a horse. But they did take with them to Asia their woolly sheep, their wheat grains, neither of which is indigenous to the area, and their textile skills. Another mummy is buried with a winnowing basket still containing wheat [Barber, 1999, p150, p37, p74, p76]. Those found at Qizilchoqa (Wupu) near Hami, more incredibly, still had intact multi-colored woolen clothes, and in some cases even appeared to be wearing tartan [Barber, 1999, p133, p138]!

Elizabeth Wayland Barber was particularly impressed by their skilful use of textiles, especially given that one key innovation from acquiring the use of wool was the making of felt. She found examples of ornamental appliqué spirals sown onto felt and evidence of a twill tapestry coat made with a sophisticated weave. Interestingly, both the lozenge and the spiral motifs appear on items of clothing, and spirals were also painted in yellow ochre on the faces of some of the mummies. Not only have relatively complex garments and the fabrics themselves survived, but so have extremely vivid colors – brilliant blues and reds – with tunics made of brightly colored felt edged in piping of a contrasting color. The ancients clearly knew how to make wool white because naturally pigmented wool will not hold colored dyes [Barber, 1999, p39].

Celts go West

While the Celts that became Tocharians are likely to have gone south through the Caucasus before they went east, the Celts that went west chose a more northerly route. After 5600 BC, when the Euxine Lake (cf the Ice Map) had flooded to form the Black Sea, a southern route to Western Europe became less attractive [Fagan, 2004, p111]. The Celts' preference for the northern route is indicated in the many Celtic toponyms found in the region, such as the Danube, which includes in its name the Celtic word for water, *danu*.

Other toponyms, such as those relating to the Celts' interest in salt and their long involvement in salt mining, point to their presence in central Europe. The Celtic word *hal* or *gal* for 'salt' could be at the root of their description as 'gaelic' peoples. Thus, the salt mines in Austria have names like Hallstat and Hallein. 'Gal' appears as part of regional names as far apart as Poland and the northwestern coast of Spain, 'Galicia'. The presence of the Galatians in Turkey at the time of St Paul is a possible indicator of a much earlier influx of people who had invaded northwest Anatolia through Troy from the Balkans, around 1200 BC, and who may well have been Celts.

What Happened to Civilization?

There is no doubt that whatever happened in 3195 BC was a frightening experience – no wonder the Celts feared the sky falling on their heads. So, what happened to civilization? Did it survive and, if so, how did it survive? The Indo-Europeans on the periphery may have absorbed many of their concepts before they left their common base, but they were not driven to building cities. The Celts, for instance, preferred to live in fortified hilltop dwellings. Those who went east and became the Tocharians similarly did not create cities on the western fringe of China.

From the perspective of our current environmental concerns, it is of great relevance to consider the impact of this natural disaster on civilization, which at that time was mostly to be found in Mesopotamia. Did the civilizers move and, if so, were their concepts well enough understood to survive being transferred to other countries over such long distances?

Chapter 8

What Happened to Civilization after the Fourth Millennium BC Disaster?

It was not just the Indo-Europeans who decided to move when the sky 'fell down' in 3195 BC: the city builders relocated too. Even though civilization did re-establish itself in Mesopotamia, there was a general exodus down the trade routes. What seems to have happened is that Mesopotamians left the country through the Persian Gulf to the south and developed first more of an entrepôt on the island of Bahrain. From there there was a positive explosion of civilization linking Mesopotamia to the Indus valley in the east (referred to in ancient texts as *Meluhha*); while in the west, some went to Lebanon and Crete, and, others on crossing the Red Sea, invaded Egypt through the Eastern Desert, thus stimulating the flowering of one of the greatest examples of ancient civilization that the world has seen[Rohl, 1998, p314].

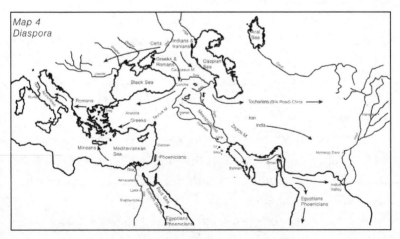

Map 4 The 4th Millennium BC Diaspora

We know that civilization was eventually restored in Mesopotamia, partly from the records left in ancient cuneiform tablets that date to the third millennium BC, although the initial decision to leave Mesopotamia may have been due to internal pressures within that country. Archaeologists have noticed that Semitic tribes referred to as 'proto-Akkadians', from Syria and northern Mesopotamia, began to come south in waves.

These tribes created a new kingdom on the north of the alluvial plain in a province interestingly called *Uri* by the Sumerians in the south, but otherwise described as the 'Kish civilization'. The Kish civilization was a 'single political configuration with shared language, culture and writing extending from Babylonia to western Syria'. It is thought that their arrival might have contributed to the collapse of the 'fragile Urukian commercial network' that existed in the south – Uruk being one of the great Mesopotamian cities [Maisels, 1999, p124]. Certainly, an archaeological layer that denotes a cultural change, such as a radically different style of pottery, seems to have started at Uruk around 3100 BC, before ending abruptly 300 years later in 2800 BC [Rohl, 1998, p181].

The Exodus – East and West

There is doubt about when precisely the civilization model arrived in the Indus valley. Its appearance there in the third millennium BC was relatively brief, lasting no more than about 1,000 years, with its period of greatest impact between 2600 BC (about the time of the building of the Great Pyramid in Egypt) and 1800 BC.

There had been some early influences of civilization in the Indus valley, similar to those in the Golden Crescent, as far back as c.7000 BC. At one site, Mehrgarh, there were rectangular mud-brick buildings and evidence of domesticated wheat, barley and goats from the west, as well as domesticated local zebu. Mehrgarh had no pottery but stone vessels. Seashells from the

Arabian coast were found, together with lapis lazuli from Afghanistan, implying that Mehrgarh was an important part of the trading arrangement involving the lapis lazuli [Mithen, 2004, pp408-410].

During the third millennium BC, however, the Indus valley witnessed the introduction of a more comprehensive interpretation of the model of civilization. Improvements were made to basic techniques like pottery manufacture when the locals were introduced to a new firing technique which they still use today. Its cities were laid out in the straight lines of the gridiron pattern, and examples of plumb bobs and surveyors' sighting devices are on display in the Delhi National Museum. Places like Harappa or Mohenjo-daro had evidence of temple areas within citadels or 'acro-sanctums' in the high parts of cities. These temples had platforms that were used as fire-altars, large tanks lined with bitumen for bathing rituals, and grain stores [Maisels, 1999, pp223-4].

Where the Indus valley differed slightly from other examples of the model was in its apparent openness, peacefulness and greater equality. The citadels which contained the features of sacred ritual – the altars, the bathing tanks – had no strategic value and did not have the restricted access of similar sites in places like Egypt or Mesopotamia. Writer C K Maisels describes them as 'community ritual precincts'.

There is also little evidence of armies, and the gateways at Harappa 'are not defensive structures'. Another curiosity is that copper and bronze would appear to have been made available to all and not just the temple or the most important households [Maisels, 1999, pp222-4]. Maybe it was these differences that made the Indus valley civilization vulnerable to attack, and possibly why it disappeared – although J P Mallory is of the opinion that this civilization 'did not so much collapse as devolve into another configuration' [Mallory, 1991, p249].

What arguably changed for the Indus valley civilization was

the arrival of northwestern invaders – probably our Indo-European relations – who were more interested in trading cattle and fighting wars than in living in cities and the finer points of urban life. These Indo-Europeans had begun to move over 1,200 years earlier, and by 1800 BC some of them arrived in India. They brought with them a different language and culture. Dravidian was the language that had previously dominated the Indian subcontinent; the continued presence of Dravidian loan words in Sanskrit indicates the later intrusion of this latter language into India [Mallory, 1991, p44].

Some indication of the direction from which these invaders came can be found in the oldest texts in Indo-European Sanskrit, the *Vedas*. Although they were not written down until the 6th century BC, they were written in an archaic language and their roots lie much earlier. Mallory describes the society in the hymns as bearing little similarity to the urban sophistication of the Indus valley civilization. He calls it an 'illiterate, non-urban, non-maritime' society, basically 'uninterested in exchange other than that involving cattle', and 'lacking in any forms of political complexity beyond that of a king whose primary function seems to be concerned with warfare and ritual'.[8] These hymns thus convey a cultural and geographical world that suggests their composition around 1500-1200 BC, reflecting a simpler life as it might have been lived on the Bronze Age steppes far to the north of India. There is also reference in the hymns to the destruction of citadels with horses and chariots – the use of horses being one of the identifiers of Indo-Europeans [Mallory, 1991, p37, p45].

Bahrain the Entrepôt and the Phoenicians

Until c.1800 BC the Indus valley was therefore part of the trade network based around Bahrain.[9] Like the Indus valley, the main settlement period on Bahrain occurred around 2600 BC, declining around 1500 BC – possibly because of a loss of trade with the Indus valley. Bahrain had been occupied during the Neolithic period and had long been of importance to the Mesopotamians. A 1954 excavation on Bahrain found temple seals and other objects similar to ones that existed in Ur [Baigent, 1994, pp27-28]. In addition, both Bahrain and Mesopotamia shared the same god Enki [Rohl, 1998, p232].

The link between the Phoenicians and Bahrain is, however, a little more unexpected. The Mesopotamians who went to Bahrain may have encountered a separate group of people on that island, before traveling on to Egypt. This group, which was to become very useful to the Mesopotamians in their new home of Egypt, then relocated to the Lebanese coast (Canaan). It was here that they became recognized as the Phoenicians, the people who took on the role of traders for the Egyptians.

Archaeologist David Rohl makes a number of points in support of this contention. Rohl claims that the Lebanese themselves believe that 'the ancestors of the Phoenicians came from far-off Bahrain'; a belief shared by classical writers (e.g. Justin, Pliny, Ptolemy, Strabo and Herodotus). Rohl says that the two islands of Bahrain were originally called Aradus and Tylos and that these names were then re-used for the two Phoenician cities Arad and Tyre [Rohl, 1998, p305].

When the Phoenicians suddenly appeared on the eastern edge of the Mediterranean some time after the beginning of the third millennium BC they brought with them their culture, which included metalwork and the potter's wheel, and their role in life already established. They created the great trading cities of Tyre, Sidon, Ugarit and Byblos (Byblos was a Greek name – the original Phoenician name was Gebal, Gubla, or Gubal). They did not call themselves Phoenicians: that was a later Greek name based on the Greek word for 'purple' (*phoinikes*), referring to the famous dye that the Phoenicians obtained from the shell of the murex [Aubet, 1993, p5, p7]. Their own name was *can'ani* – Canaanite, meaning 'merchant'[10] – which precisely described the function they performed for the Egyptians who, from the beginning, were not interested in being merchant seamen. One of the earliest Egyptian papyri, dating from about 3000 BC, refers to 40 Phoenician merchant ships with cedar of Lebanon for Egypt. Another, dating to about 2600 BC, concerns a shipment of wood and oil for Egypt from the Phoenician city of Byblos [Aubet, 1993, p146, p18].

Minoan Crete

The other key partner in this Mediterranean trading equation between Phoenicia and Egypt was Minoan Crete. Although we are accustomed to thinking of Crete as part of Greece, that was not the case in ancient times. The evidence for contact between Crete and the Greek mainland is 'very slight throughout the early Bronze Age' [Fitton, 2002, p39, p63]. It was only later when Mycenaean warriors took an interest in Crete, sometime around the middle of the second millennium BC, that the link between the Greek mainland and the island became more strongly established.[11] Crete had been occupied since at least 7000 BC when the earliest settlers 'used types of bread wheat which are attested in Anatolia but not, as far as we know, on the Greek mainland at this time', which is a strong clue as to the direction they came from.

But it was from around 3000 BC that archaeologists have noticed that Crete experienced a 'rapid surge forwards into the Bronze Age', with changes in metal technology. Suddenly, domed 'tombs', the *tholoi*, begin to appear. Motifs like the snake goddess, the moon-sickle and rayed sun-disc also make their appearance, as well as the double axe, arguably the 'religious symbol most strongly associated with the Minoan culture' [Castleden, 1993, p29, p153, p155, p129, p134].

The Secondary Products Revolution in agriculture now arrived with the use of the plough, new crops of olives and grapes, and the creation of specific places for the processing of wool. Like the Phoenicians, the Minoans became known for the use of the murex shell to produce purple dye. Historian J L Fitton even describes the Minoans as being the 'forerunners in this prestige industry' [Fitton, 2002, p41, p20].

Just as in the Indus valley, the Minoans had their temple-palace complexes at the centre of urban life – the most famous example being that at Knossos which contained granaries and workshops. Knossos, in the form that we now know it, dates from the height of the Minoan period toward the end of the third millennium BC, although there was earlier building on the same site. From that later time, there is also evidence of sacred sites in high places which – again, like the Indus valley – left traces of walled areas and fire altars with 'extensive ash deposits'. The sophistication of Crete was echoed on the smaller island of Santorini to the north, where volcanic ash has protected evidence of two- or three-storey houses, ashlar masonry and an efficient drainage system dating from the time of the Minoans [Fitton, 2002, pp58-60, p168, p171].

From the start of the third millennium BC, the beginning of the Cretan Bronze Age, all the evidence suggests that the focus of trade activity lay between Crete, the Phoenician coast and Egypt [Fitton, 2002, p63]. Crete was even given a similar sounding name from these directions. There are references on Theban tombs dating to around 1470-1450 BC to the Minoans as 'Princes of the Land of Keftiu' [Castleden, 1993, p12]; the Bible called Crete *Caphtor* which is likely to have been its Canaanite name; while 18th century BC documents found at Mari on the Euphrates mention *Kaptara*. One undeniable reason for the importance of Crete is that the prevailing winds in the Mediterranean are in an anticlockwise direction so that the Phoenicians were obliged to sail via Crete to make their deliveries to Egypt [Fitton, 2002, p64].

Mesopotamians invade Egypt

David Rohl is convinced that the people who invaded Egypt at the beginning of the third millennium BC were Mesopotamians who entered the country from the east. He has identified two

In addition the style of the Heb Sed festival court in Saqqara Egypt
resembles the mudhifs of the Marsh Arabs in Southern Iraq

plant types in decorative motifs in early Egyptian architecture
linking Egypt to the area north of Mesopotamia, the Golden
Crescent. One plant is the Madonna Lily, and the other is the
Giant Hogweed. Both these plants he considers to be out of place
in Egypt. The Madonna Lily is not native to Egypt, although it
was grown there for its perfume. Both this lily and hogweed have
as their natural habitat the Crimea, the Caucasus from the Black
Sea to the Caspian Sea, and parts of northern Turkey [Rohl, 1998,
p379, pp381-382].

He points out that the heraldic plant of Upper Egypt used in
carvings at Karnak is surprisingly not the lotus but the Madonna
Lily; while Rohl regards the Hogweed as the inspiration for the
Doric-style fluted columns in the earliest First Dynasty tombs at
Abydos. Although these columns are clearly representative of
some sort of plant, they are not based on Egyptian papyrus, as
papyrus does not grow straight and lacks the carved petioles of
the Egyptian columns. The other standard explanation for the
Doric-style columns - bundles of reeds - does not make sense as
the fluting on the columns is concave, not convex. The most likely
contender is the Giant Hogweed, *Heracleum Giganteum*, a native
of eastern Anatolia and the Caucasus, which grows to 5-6m in
height, with the strength of bamboo when dry. It has no equiv-

alent in Egypt or North Africa [Rohl, 1998, pp376-7, pp374-5].

Rohl is convinced that the Mesopotamians arrived through the Eastern Desert, because of strange rock art at *wadis* on the major trans-desert routes between the Nile valley and the Red Sea – routes which also led to the Eastern Desert gold mines [Rohl, 1998, p275]. These primitive images, scratched onto rocks in the middle of the desert, show images of boats like the one that was buried at Giza by the Great Pyramid, carrying individuals, possibly deities, with their arms uplifted or with a pair of plumes on their heads, in similar images to those found in Mesopotamia.

When a much later 13th century BC pharaoh, Seti I, had a new temple cut out of the rock, the Temple of Kanais, in the *wadi* east of Edfu, he made it quite clear that he was dedicating this temple to Amun-Re 'and his divine ennead' to celebrate the 'reopening of an ancient route to the Red Sea and Eastern Desert gold mines' [Rohl, 1998, p275].

Trade routes through the Eastern Desert continued to be important. There are many references in Egyptian texts to the quasi-mythical Land of Punt, or 'God's Land' (*Ta-Netjer*), referred to as being located 'on both sides', which could have meant both sides of the Red Sea, of Arabia or perhaps of the Horn of Africa. Certainly, the Egyptians were keen on the frankincense and myrrh that came from ancient mountain terraces in Yemen, as well as the more exotic African trade that traveled up the southern Nile. There are beautiful illustrations at Deir el-Bahri of expeditions for the Egyptian queen Hatshepsut showing ivory, monkeys, cheetah and panther skins, and plants, as well as curious images of 'asiatic' traders with pointed beards and their African wives.

Safety

Thus, by the early third millennium BC, Egypt had become the centre of civilization with all the other destinations – Minoan

Crete, Bahrain, the Indus valley, Canaan and even Mesopotamia to some extent – in effect as outposts or satellites. It is from this time onward that Egypt becomes so strongly associated with every aspect of the model. There was one good reason why Egypt gained such prominence: at that time it was considered to be one of the safest places in the world.

As mentioned in the previous chapter, when the comet hit the Earth and destabilized everything, there was a longitudinal 'line of neutrality' where there was no change in sea level, stretching from Scandinavia, through Italy and the centre of Africa – Egypt [Dunbavin, 1995, p76-79]. Safety had to be the attraction because, after all, the Egyptian climate was not particularly appealing. The catastrophe had as bad an effect on the climate in Egypt as anywhere else and caused a dramatic change in habitat.

Geologist Robert Schoch and others confirm that Egypt had had a wet climate from between 10000 and 8000 BC – the Nabtian Pluvial period – which finally ended in 3000 BC [Schoch, 1999, p40]. Rainfall in East Africa had been annually 150-440mm, and there were many freshwater lakes in the desert which lasted until 2550 BC, when they finally dried up. Rock art in the Egyptian Eastern Desert, which dates from 4000 BC or earlier, plainly shows images of elephants and giraffes; so wet and green was it then [Fagan, 2004, pp151-2].

The newly arrived Mesopotamians took on a country at the end of the fourth millennium which was becoming increasingly drier. This created the desert which we now associate with Egypt, although that was not a complete deterrent, as the Nile, with its annual floods, provided plenty of fertility. It is this evidence of climate change which has created so much suspicion around the true dating of one of Egypt's most famous monuments, the enigmatic Sphinx.

The Pre-Sand Sphinx
Conventionally the Sphinx is dated to around the middle of the

third millennium BC, the time of the pharaoh Khafre – known to the Greeks as Chephren – partly because of the likeness of its face to that of the pharaoh. But dating the head of the Sphinx on the basis of the resemblance of its face to a particular pharaoh is unconvincing when the state of the rest of the body is not taken into account. In any case it is highly probable that the head of the Sphinx was re-carved at some point – maybe at the time of Khafre – because it is proportionally smaller than the rest of the body [Schoch, 1999, p44].

Boston University geologist Robert Schoch has closely examined the monument, coming to the conclusion that it must be older than is commonly thought. While Prof Schoch makes his observations from a geologist's point of view, he is not alone in believing the Sphinx to be older. The two well-known Victorian Egyptologists, Flinders Petrie and Wallis Budge, also took the same view. According to a more recent Egyptologist, Selim Hassan, literary references to the Sphinx from New Kingdom to Roman times 'all considered the Sphinx older than the pyramids' [Schoch, 1999, p45].

First of all, Prof Schoch looked at how the Sphinx was made. It was excavated out of the living rock and so sits, in a sense, in its own quarry; only the head and upper back are exposed above ground level. The limestone from the quarry was then used to construct the Sphinx and Valley Temples. The weathering pattern on the core walls in the temples, the sides of the quarry and the Sphinx thus all date from the same time. All of them show the same deep vertical fissures. The Valley and Sphinx Temples were repaired later, possibly during the time of Khafre, as they have inscriptions dating from then on the repairs done with granite ashlars from Aswan, cut perfectly to fit the eroded limestone underneath. The Sphinx was also repaired during the Old Kingdom [Collins, 1998, p11; Schoch, 1999, p36, p41].

Prof Schoch compared the Sphinx erosion with that of the Saqqara *mastabas*, which are dated no earlier than 2600 BC and show much less erosion in spite of being 'fragile mud-brick tombs', indicating that they could be of more recent construction than the Sphinx [Schoch, 1999, p36, p39]. Next, Prof Schoch took into account the fact that the Sphinx has spent a lot of its existence with its lower half protected, as sand blew in and filled the quarry up to its neck. As a result, the weathering pattern is uneven on the Sphinx. But, in Prof Schoch's opinion, the pattern seen on the Sphinx is not due primarily to wind-driven sand but to *water*, which makes little sense when surrounded by desert sand, even though the Nile is nearby.

If the Nile had flooded the Sphinx enclosure, however, as writer Robert Temple believes [Temple, 1999, p25, p32], then there would even have been a pattern around the lower body and especially on the front paws. The erosion is, however, 'most pronounced on the much higher back'. Prof Schoch carried out seismology tests with Thomas L Dobecki in April 1991 that showed that the enclosure surrounding the Sphinx to be unevenly weathered [Schoch, 1999, p48, p38]. While the front and both sides of the Sphinx are deeply weathered to a depth of 6-8ft below the level of the enclosure's currently exposed surface, at

the west end at the back of the Sphinx the weathering is much less, to a depth of only 4ft.

Schoch thus came to the conclusion that these deep vertical cracks in the upper body of the Sphinx, and in the surrounding enclosure and temples, could only have happened at a time when

The Sphinx with the stela at its paws

Egypt was exposed to long periods of intense rainfall. As we know, the Nabtian Pluvial began sometime between the eleventh and ninth millennia BC [Schoch, 1999, p40] and so, for the rainfall to have had such an impact on the limestone, the Sphinx must have existed at a time when the Sahara was wetter, after 8000 BC and before 3195 BC [Fagan, 2004, p158]. The Sphinx must therefore date to what the Armenian philosopher Gurdjieff referred to as pre-sand Egypt [Gilbert, 1996, p114]. Ever since, the Sphinx has sat, enigmatically, facing the rising Sun for thousands of years: as the *stela* placed between its paws states, it has remained the 'presider over… the splendid place of the First Time' (known in Egyptian as *Tep Zepi*) [Picknett & Prince, 1999, p51].

Routes into Egypt

Corroboration of early activity in Egypt comes from an archaeological site at Maadi, across the Nile from the Giza plateau, which is on a trade route to the east linking with the copper mines in Sinai and to the north with the delta for shipping [Baigent, 1998, p176]. When Maadi was excavated in the 1930s it was found to have early traces of metalworking, unusual underground dwellings and 'huge, burial store-jars… alien to Egyptian practices', evidence of a wide range of cereals and the earliest remains of a donkey anywhere in Africa [Maisels, 1999, p52]. The jars are of hard materials that are difficult to work with, like basalt, granite and diorite, in addition to limestone and alabaster [Baigent, 1998, p177]. Most of the dates that can be established for Maadi are clustered around 3650 BC – at the time that the Ubaid were in Mesopotamia [Maisels, 1999, p51].

What makes Maadi especially interesting is its possible links with Syria and Palestine, which mean that it could be even older. Its cellars are similar to shallow underground dwellings found at Jericho, dating to around 5000 BC. Jericho had a stone wall around its town and a 30ft high stone tower since 7000 BC [Baigent 1998, p179]. It is in this period, 5000-7000 BC, that the civilizers had their influence on Çatal Hüyük in Turkey and elsewhere in the Golden Crescent.

It is therefore possible that perhaps some time at the end of the sixth millennium BC, the civilizers put down, as it were, a marker on the Giza Plateau, creating the Sphinx and the nearby Valley and Sphinx Temples. The civilizers who arrived at the end of the fourth millennium BC were in effect returning to Egypt, although this time they took a different route through the Eastern Desert, invading Egypt from the Red Sea after leaving Bahrain.

Even if this was some sort of return, how did the ancients retain so much information about civilization in a pre-literate society, to the extent that they were able to reinstate so success-fully the archetype of civilization in Egypt? What was the secret of the ancients' detailed knowledge regarding civilization? How did they know how to accomplish so many similar feats when in theory they were not as 'civilized' as we are? They didn't have access to the sort of technical manuals that we take for granted. They didn't even necessarily know how to read and write. Yet clearly they were able to re-establish the same complex pattern of civilization regardless of location, and on a significant scale. They must have had an alternative approach to accessing information – a method that is not familiar to us.

Chapter 9

Metaphysics and Knowledge

It is our own cultural problem that we regard societies without writing as being primitive. For us, one of the most significant characteristics of civilization was the development of writing. We think of knowledge as being stored in books – or today, computers. We don't really consider civilization to happen properly until the invention of an alphabet that we can understand. We are, however, missing the point. We are in danger of confusing two separate developments – the writing down of knowledge, and the writing of administrative records and business contracts – in our rather pointless search for the origin of writing. In ancient times there were much clearer distinctions.

The Sumerians were particularly keen on administrative records for their cities. Their populations could run into tens of thousands of people, requiring a significant degree of organization. Writer Charles Maisels comments that Mesopotamian texts written in cuneiform from Lagash, dating from the Ur III period, toward the end of the fourth millennium BC, showed 'high managerialism of an order that did not reappear until the 20th century AD' [Maisels, 1999, p167]. Other Mesopotamian texts from Uruk and Shuruppak confirm the highly organized nature of Mesopotamian life.

The Shuruppak texts, known as the Fara Tablets, give a particularly interesting insight into how life was organized. One Fara archive 'refers to men and officials coming from other Sumerian cities, as well as a particular group of people called *gurus*, who came to Shuruppak for certain services. The large numbers of people in these groups, ranging from 1,300 to 160,000, are remarkable. The *gurus* seem to have been counted in units composed of 680 men and could be assigned to different officials. While some of the tasks they performed seem to have been agricultural, a number of texts specifically mention that they were military'.

On the basis of these very old texts we credit the Sumerians with the invention of writing at the end of the fourth millennium BC. But, as historian Gwendolyn Leick states, the Uruk texts show 'complex numerical data and administrative responsibility' (lots of lists and book-keeping), not literacy as such [Leick, 2001, p65]. Their version of cuneiform, with over 2,000 separate symbols [Knight & Lomas, 2000, p220], did not evolve into any other form of writing. This lack of evolution is apparent in that pictograms (another kind of cuneiform) are still in use in Japan and China today. Cuneiform also did not evolve into Egyptian hiero-glyphics, in spite of both sharing some of the same symbols.

Whereas the Sumerians had relied on making dents in wet clay tablets to form symbols or pictograms, the Egyptians did not even bother with wet clay. From the beginning they used papyrus and ink which, as David Rohl points out, makes it easier to represent abstract images in hieroglyphs [Rohl, 1998, p319]. According to Rohl, the ivory labels found near Abydos are the 'first examples of hieroglyphics in Egypt and yet they appear for the first time in their fully developed form'. He comments on how 'remarkable this sudden appearance of writing is in Upper Egypt' [Rohl, 1998, p354].

William Arnett wistfully clings to the hope that 'if enough pre-dynastic inscribed material can be collected and catalogued, it may be possible to write an earlier chapter to the history of Egypt and to cast light upon an epoch which remains obscured in what, at present, must be termed prehistory', because 'it cannot be ruled out that we may yet see the origin of hieroglyphic writing' [Arnett quoted in Rudgley, 1998, p16].

The Art of Memory

It is misleading to talk about the arrival of writing as a break-through in human progress. Writing didn't exist, not because the ancients were primitive, but because they didn't necessarily *want* it. What mattered to the ancients was the art of memory – and this remained of importance to the Celtic Druids long after everyone else had taken up the pen, even though the Druids had a script they could use – the Ogham alphabet. The Celtic druidic schools preferred to teach students to memorize information and continued to do so after the Romans had invaded Britain, bringing with them their more 'advanced' literary skills. It is thought that the name Celt may derive from an Irish word meaning 'secret' or 'hidden' which refers to the way they trans-mitted their knowledge [Berresford Ellis 1990, p9].

Plato's *Phaedrus* has a story that Socrates tells, illustrating this attitude. In the story Plato is openly critical of Thoth, the Egyptian god credited with inventing the art of writing, when he is presenting his invention to the king at Thebes. The king is not happy, on the basis that 'this invention will produce forget-fulness in the minds of those who learn to use it, because they will not practice their memory'. The king fears that those who learn to write will give the 'appearance of wisdom, not true wisdom, for they will read many things without instruction and will therefore seem to know many things, when they are for the most part ignorant' [Yates, 1966, p52].

On the one hand, the priests either memorized or jealously guarded their written records and symbols with layers of meaning as well as individual words. Commemorative *stelae* of military victories, or diplomatic exchanges between kings, are perhaps the exception to the predominant role of the priestly scribe working on the more arcane aspects of temple work. The Egyptians did have a hieratic, more cursive form of writing for everyday use, as well as hieroglyphs.

On the other hand, there is no doubt that Thoth was right:

writing is a useful and necessary skill for an advanced society. T J Cornell, in discussing a much later period, namely Rome in the 1st century BC, acknowledges that literacy is not a prerequisite for urban society, but it made it possible for the Romans to 'reorganize and reclassify' masses of complex data and enabled the Roman city state to divide and subdivide its citizens into different kinds of functional groups – especially for the conduct of warfare [Cornell, 1995, p104].

It is significant that our alphabet, the Roman alphabet, evolved neither from the Sumerians nor the Egyptians but out of the practical needs of traders, the Phoenicians. The Egyptians had, anyway, been a closed country to most, apart from the Phoenicians and limited contact with the Greeks from the 6th century BC. After the 4th century AD even the Egyptians lost the ability to understand hieroglyphs.

What we forget is that today we use one system of writing for all our purposes, mundane or otherwise – with one exception, the sciences. Mathematics, physics and chemistry all still use symbols in the form of either the Greek alphabet or other sigils that have come down to us over the years. Such a complex form of written language based on symbols is bound to be limited in its use to those people (such as priests or modern-day scientists) who are specially trained to understand it. Indeed, the word *hieroglyph* means 'mark of the priest'. But the drawback of symbols is that they can have so many meanings [Knight & Lomas, 2000, p222]. It would have been useless for the Phoenicians to have used hieroglyphics in the compiling of business contracts because the last thing they needed was a multiplicity of meanings.

Magic and Science

The kind of knowledge that the priesthood retained regarding the civilization project could not necessarily be communicated through writing. It was accessed through a quasi-spiritual experience. There was no distinction in the ancient mind between science, religion or magic. Much of what the ancient priesthood knew was what we might call science and what they called magic. But although the ancients had scientific knowledge, they were not 'scientific'. The ancients had no notion of *objectivity*, of being 'outside' knowledge.

For the ancients, the human being was fundamental to the process of obtaining knowledge: who was making the enquiry was as important as the question being asked. The priest or priestess was a living repository of information. Only they knew the techniques for finding out more – techniques which did not involve scientific experimentation but rather involved what we might regard as a mystical experience that to them was magical. Their experience of knowing was essentially a transcendental, spiritual experience, best summed up in the Greek word *gnosis*. *Gnosis*, meaning knowledge, has become associated with early Christian sects known as Gnostics, as it has distinctly spiritual overtones, but it applies equally to more mundane knowledge as well.

In ancient times, magic was serious. The Egyptians in particular had notions of different levels of magic. Magical tricks were the lowest form of magic, called *goetia*, used to demonstrate the power of magic. With *goetia* the divine force animates an object through imprinting. Then there was *magia* in which the presence of the gods or divine force was experienced only through some material object or medium. Finally, the highest level, *theurgy*, was a form of magic where the gods were present to a person who might be awake, dreaming or in an ecstatic state [Lindsay, 1970, p207].

In Egyptian temples an important role belonged to the magician or master, the *Ur Hekau* (or *Ur-t Hekau*, the female *Hekau*). In the Pyramid Texts he took the form of a serpent in the *Duat* (the shadow world), later anthropomorphized into a man holding two serpents in each hand. He was known as 'Lord of

Oracles and Revelations and He Who Predicts What Will Happen'. The *Ur-Hekau's* primary function was the 'transmission of *heka* from the sanctuary to the other priests, the Royal House, and in the healing temple, to those deprived of it through illness'. The *heka* – from which we get the word 'hex' which to us means curse – was neutral, neither black nor white. It was a mystical force, a catalyst that uses the 'transformative power of individuals, places and the *Neteru*' [the gods] to preserve cosmic order and to maintain the divine plan [Clark, 2000, pp358-9].

In spite of the *Ur Hekau* having the title 'He Who Predicts', there was not necessarily any sense of fatalism in his engagement with the divine plan. Instead, there was an idea of a dynamic dialogue between the magician, magus or priest and the spirit world. The Hermetic dictum, 'as above, so below', could be reversed, if one were initiated in the techniques and knew how to operate sympathetic magic, the principle of correspondences – that 'like attracts like'. Thus, the *Ur Hekau* could effect change by magical manipulation of one substance that corresponded with another [Clark, 2000, p360]. Nowhere is this principle of dynamic interaction better illustrated than in the ancients' approach to one form of divination, horary astrology.

Fatalism *versus* Freewill

Astrology was an important subject for the ancients. Groups of priests, who spent their time observing the night sky, were ready to provide the king with information regarding any important omens that might affect his actions. One group in particular, the Chaldeans or the *Chaldini* [Gilbert, 1996, p288], retained a powerful monopoly in this field in the Middle East for a long time. They were part of the network of wise men in the Middle East who helped to create the libraries of Babylonia, Assyria and the Hittites and who, toward the end of the second millennium BC, had developed astrology into more of a scholastic discipline [Baigent, 1994, p66]. They survived as a separate ethnic group, a priestly caste, for thousands of years, with their own secret language [Velikovsky, 1978, p172], taking up residence in a number of places in the Middle East, from the northern Euphrates to southern Babylon.

Ancient horary astrology was quite different from the natal astrology with which we are familiar today. It wasn't the timing of an event – such as the time of a birth – which mattered with horary, but the precise moment at which the one casting the horoscope, the astrologer, made the decision to ask the question on behalf of the querent [Cornelius, 1994, p106]. The querent may even have had the question in mind for a number of years [Barclay, 1990, p28].

The Greeks referred to this moment as the *katarche*, meaning 'beginning' or 'initiative' (as in Homer's *Odyssey*, when *katarche* is used in the context of 'to begin the rites of sacrifice') [Cornelius 1994, p138; Barclay, 1990, p135]. To make horary was itself 'as if to begin the rite of sacrifice' [Barclay, 1990, p150]. As writer Geoffrey Cornelius states, 'The use of divination 'authorized' a course of action at its inception, by giving it the sanction of the gods. To take an auspice was to undertake human initiative under the guidance of the gods'. To inaugurate something thus became a *'ritual* inception of a major matter' [Barclay, 1990, p142-3].

Olivia Barclay, a modern astrologer, explains that 'The moment of the question is a moment of contact with a greater intelligence. To that extent it is divine. The planetary and fixed star positions then extant are an expression of that greater intelligence, a writing that our pea-sized brains can decipher, once we understand the code' [Barclay, 1990, p27]. The greater intelligence that she refers to was traditionally regarded as 'intelligences between mortals and gods' [Cornelius, 1994, p141], intermediaries, referred to in a variety of guises, from guardian angels to genie or *djinn*. Perhaps the most interesting term, now a pejorative word, for these intermediaries of the spirit world is an old Greek word, *daemon*. The Greek term for a sign or omen is also *daemonion*, meaning 'the divine thing' [Cornelius, 1994, p191]. In this context the term a *daemon* is entirely benign.

A much later 12th century AD astrologer, Guido Bonatus, who learnt about horary astrology from Arabic sources, made it clear that enquiries could not be made 'on trifling occasions, or light sudden emotions, much less in matters base or unlawful'. The astrologer needed to fully understand the question being asked and ask it with serious intent. According to Cornelius, seriousness of motive was necessary to 'guarantee "radical intention" and thus "radicality" in the symbolism and the judgment' [Cornelius, 1994, p117].

This early astrology that one finds in Mesopotamia was not concerned with the individual – there were no birth charts – and has no references to the zodiac. It was only concerned with omens affecting the king and the state [Baigent, 1994, p43]. Fundamentally, the astrologer was not looking for a specific answer in the manner of a prediction, but to unravel knots and make decisions easier when faced with a dilemma. It was a form of decision-making, setting out the king's options and letting him know what consequences might follow from choosing one course of action over another. There was an acknowledgement that the stars are 'not compelling' (*non cogunt*) [Baigent, 1994, pp154-5].

In the words of a present-day American astrologer, 'The chart merely maps a potential which can be used as the basis for constructive counseling'. It is then up to the querent what they do with the information [Doane, 1994, p7]. The person asking the question at least has a clearer picture whether or not what they are planning is a good idea. For that reason standard texts in Mesopotamia often began with 'if' [Baigent, 1994, p42], in effect using 'an astrological omen to indicate the will of the gods – and thus the outcome in terms of good or ill fortune – with respect to human initiative brought before those gods' [Cornelius, 1994, p150].

The Zodiac

How astrology and the zodiac became confused with fatalism is explained in passages of the *Hermetica*. The *Hermetica* states that, when Atum created 'this beautifully ordered universe' and installed humankind on Earth, the other gods protested and claimed that human beings would effectively ruin the Earth. Atum replied, 'I will build the zodiac, a secret mechanism in the stars, linked to unerring and inevitable fate. The lives of men, from birth to final destruction, shall be controlled by the hidden workings of this mechanism' [Freke & Gandy, 1997, pp99-100]. The *Hermetica* thus confirms that 'few can escape their fate or guard against the terrible influence of the zodiac – for the stars are the instruments of destiny, which bring all things to pass in the world of men'.

Yet the same section of the *Hermetica* goes on to say that there are some for whom 'the workings of these gods are as nothing' [Freke & Gandy, 1997, p101]. In other words, there is the possibility of free will. For these few the means of escaping fate is through having an illuminated mind, or what the Egyptians called *nous*: 'For if it's not possible to escape the condition of change any more than that of birth, he who possesses *nous* has the power of getting away from evil' [Hermes to Tat, *On the Common Intellect*, quoted in Lindsay, 1970, pp301-2]. It is 'men *without* intellect' who are 'simple puppets in fate's procession' [Zosimos, quoted in Lindsay, 1970, p326]. Astrology – particularly as horary astrology – was thus available to those with *nous*, in particular to the ancient astrologer-priests.

The other characteristic of horary, and indeed in the ancients' approach to all matters regarding knowledge, was that the quality of the individual astrologer making the enquiry also mattered. There was no sense in which a horoscope could be cast 'objectively': the astrologer appeared in the horoscope as well as the querent and so was personally and actively involved [Cornelius, 1994, p109].

At the same time, the astrologer had the freedom to decide when was the right moment to make the enquiry – as the 12th century AD Bonatus said, 'By the election of his will the person does actually inquire' [Cornelius, 1994, p116]. Much also lay in the individual astrologer's skill in interpreting the horoscope. Above all, the astrologer must not be afraid to discard whatever, within the chart, does not result in clear symbolism. As Cornelius puts it, horary is beyond mere technique: the 'ability to see symbolic fittingness is far more a guide to sound judgment than the mechanical application of strict horary rules' [Cornelius, 1994, p123-4].

The Royal and the Religious

There was, however, one form of enquiry that involved such a deep level of spiritual inquiry – a form of shamanism – and was regarded as so special that it required a particular location and an elaborate rite. This was the aspect of the model of civilization that involved the critical relationship between rulers and their priests. It was this part of the model which sheds light on why the role of the ruler, the king or pharaoh, was more 'divine rite' than 'divine right'. The ruler, as the earthly representative of his people, was charged with ensuring that civilization continued, that people carried on living in 'ma'at' (in truth and harmony) in a prosperous and fertile land. To that end, the ruler underwent initiatory training and then regular rituals under the supervision of the priesthood.

As mentioned earlier, in Egypt the training of the ruler, the pharaoh, would have taken place under the guidance of the priests called the *Shemsu Hor* (or Followers of Horus) in the *Per Ur* ('the Exalted House' or 'House of Foundation'), the name of the sacred precinct at Nekheb. This place name so impressed itself on the Greek mind that the Greeks confused it with the person, hence the title 'pharaoh'.

The purpose of this training was so that he could take part in an essentially transcendental, mystical experience that was supposed to occur during an important Egyptian festival of renewal, the *Sed* or *Heb Sed* festival. The *Sed* festival was not an annual event and is often referred to as the pharaoh's 'jubilee', even though it sometimes occurred more frequently than every thirty years [Naydler, 2005, p71].

These rituals, involving the symbolic death and rebirth of the pharaoh or king, were a widespread feature of ancient civilization everywhere. These rites continued to be re-enacted for thousands of years in kingdoms of the Middle East, and later in the mystery schools of ancient Greece. Some even argue that perhaps the most famous symbolic death and resurrection, that of Jesus Christ, is in this same tradition [Freke & Gandy, 2000, p73].

In Mesopotamia, the king took on the symbolic role of the god Marduk and descended into the ziggurat, into the underworld, where he spent three days during which the god was 'confined in the mountain' so that he, as Marduk, could emerge and be 'born again'. The Mesopotamian name for this new year festival, the *Akitu* – which dates back to at least the third millennium BC and possibly before – meant 'power making the world live again' [Naydler, 2005, p56]. In Egypt, these rites were based on the legend of the murder and dismemberment of Osiris by his brother Seth. After three days, Osiris' wife, Isis, finds all the parts and restores her husband to wholeness again (with the exception of his phallus, so she creates an artificial phallus and successfully impregnates herself, later giving birth to their son Horus the falcon god). The pharaoh was expected to re-enact this legend and perform what became known as the Osirian rites on the eve of *Heb Sed* festival, which took place over five days after the annual flooding of the Nile. The purpose of these rituals was thus to connect the king to the spirit world.

In order to find out what happened in more detail we can use the Egyptian *Heb Sed* festival as an example. Public parts of the festival included the pharaoh wearing a special kilt with a bull's tail and running around a dedicated courtyard for the purposes of the festival. But it was in the secret parts of the ritual that the pharaoh performed the role of shamanic enquiry. We can learn something about these secret rituals by studying the Pyramid Texts that the Egyptians left behind on the walls of their sacred buildings; and then examine the *Heb Sed* festival itself in greater depth.

The Journey into the *Duat*

The Pyramid Texts provide many clues. The oldest of these texts – which the Egyptians regarded as having been written by Thoth – are on the walls of the pyramid of the pharaoh Unas at Saqqara, dating to about 2350 BC, although it is thought that the texts themselves are possibly copies and that the originals could have been as old as 2700 BC [Baigent, 1998, p190]. Likewise, other texts that are of interest are those to be found at the much later Ptolemaic temple at Edfu in southern Egypt, but which, again, are thought to date to much earlier times [Collins, 1998, p210].

Just as we misunderstand pyramids – thinking of them as being tombs – so we misinterpret the meaning of the texts found inside them. Because the texts refer to the journey of the pharaoh's soul in a shadow spirit world, the *Duat*, we assume that they relate to the death of the pharaoh. This mistake is not entirely surprising given that, for centuries, the Christian church has taught us that it is only after death that our soul has any importance (in that it goes either to heaven or to hell). Consequently, we commonly call these 'Coffin' texts or 'Funerary' texts, or the 'Book of the Dead' [Baigent, 1998, p191], even though the Egyptian name for these texts was the 'Book of Coming Forth by Day' [Hancock & Faiia, 1998, p68].

The Egyptians, however, had an entirely different understanding of the meaning of 'soul' from us. It is only now, as we have begun to relearn about the soul from our contact with the

Far East and India, we have become aware that a soul could have different layers. It is quite commonplace today for us to refer to a person's aura. To the Egyptians such a concept would have been totally familiar. For them the soul was a complex entity. It was not just a question of being dead, alive or being judged. Instead, the soul was composed of many layers or etheric bodies. For the Egyptians there were three main components: the *Ab*, the heart-soul or seat of integrity, the *Ka* and the *Ba*.

The *Ka* was the layer of the soul that corresponded to what we might regard as the aura, represented by the hieroglyphic image of arms with two upraised hands. The *Ka* was the vehicle transmitting vitality between spiritual and material bodies and thus considered immortal. Wallis Budge, the eminent Egyptologist, viewed the Egyptian *Ka* as being like the body's 'guardian angel' [Clark, 2000, p296].

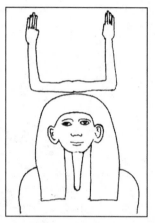

Pharaoh with *Ka* above his head

Strictly speaking, what we call the 'soul' for the Egyptians was the *Ba*, illustrated in hieroglyphics as a human-headed hawk. This aspect of the soul was the astral body, for while it needed the *Ka* in order to inhabit a body, the *Ba* could leave a body while it was still alive [Clark, 2000, p299] – for example, while the body was asleep at night. The second shrine of Tutankhamun shows the pharaoh connected with a star with his *Ba* sitting on a perch in front of him [Hancock & Faiia, 1998, p88].

Tutankhamun in front his *Ba*

Realm of Sokar

Admittedly, the impression in the Pyramid Texts of the pharaoh being dead is reinforced by a description of him descending into a subterranean chamber of the pyramid where he enters the realm of a god of death, Sokar. This deity is displayed in reliefs as a mummy with a falcon head, lying prone under the primeval mound and often shown as either reclining in a sarcophagus or on a bed with a lion head and legs [Clark, 2000, p66, p102].

The realm of Sokar

Sokar had a more complicated association with death than just being dead. He was part of a triple manifestation of the gods Ptah-Sokar-Osiris. These gods represented 'the triple powers of animation, incarnation and restoration' [Clark, 2000, p66, p103], and thus were essential to the Egyptian ideas of cycles of life and death, in terms of the soul being immortal and reincarnated in a living body. Writer Rosemary Clark describes Sokar as representing 'the latent spiritual principle within all living things, the spirit embedded in the deepest regions of matter that await arousal' [Clark, 2000, p66, p68], a description which implies the beginning of life rather than the end of it. Sokar was also linked with another god associated with death, the jackal-headed Anubis.

Shamanism

Given these textual references to the soul and its destiny, together with the discovery of texts in what appeared to be underground burial chambers in pyramids containing sarcophagi, it is understandable that modern archaeologists should conclude that what was involved was the physical death of the pharaoh. But what has been overlooked is the possibility that what was being referred to was the *near-death* of the pharaoh rather than his actual death; that the process as described occurred while the pharaoh was still alive. One of the Pyramid Text utterances even says to the pharaoh, 'You have not departed dead, you have departed alive' [Naydler, 2005, pp202-04].

On this basis, what the pharaoh was actually experiencing was an out-of-body journey – or astral-traveling. The pharaoh was behaving as a shaman, as someone who goes into a trance, perhaps with the help of an external stimulant, in order to communicate with the spirit world, usually to find the answer to particular questions.

The pharaoh was thus undertaking a journey similar to those still undertaken by shamans in hunter-gatherer societies today. For thousands of years the shamans who have lived among native peoples have adopted the attributes of a particular animal or bird – such as reindeer in Siberia or eagles in North America – climbing a symbolic ladder into the spirit world to find the answers to questions. Certainly the language of the Egyptian texts has many such shamanic images.

The Pyramid Texts frequently refer to the pharaoh taking on the form of a bird and flying up or climbing a ladder. Chapter XX of *The Book of the Dead*, for instance, talks of the pharaoh rising into the sky 'like the mighty hawk' [Lindsay, 1970, p355]. The Antechamber Texts in the Pyramid of Unas at Saqqara refer to 'a stairway to the sky [which] is set up for me that I may ascend on it to the sky, and I ascend on the smoke of the great censing. I fly up as a bird'. Elsewhere, the same texts refer to the pharaoh flying up in the form a falcon to the 'imperishable northern stars' [Naydler, 2005, p259, pp202-04].

As writer Jeremy Naydler, in his book on the Pyramid Texts, says, 'The image of the sky ladder or stairway is, along with transformation into a bird, one of the most pervasive symbols of the means of ascent from the world we normally inhabit with our ordinary consciousness to the spirit world, accessible only to visionary consciousness'. He compares the Egyptian descriptions to the mystical ladders also present in the later Orphic and Mithraic mysteries. The pharaoh was able to 'fly' because he had become a Horus-king and assumed the form of the Horus falcon, the son of Osiris and Isis [Naydler, 2005, p259, p273, p270].

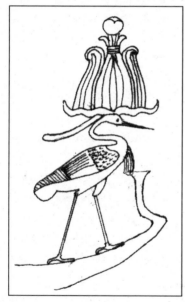

The phoenix/*bennu* bird

Equally, because the pharaoh wasn't really dead, the rituals did not concern themselves with the *Ab*, the heart-soul, but with the two other parts of the soul, the *Ka* and the *Ba*. The objective of the soul during these rituals was to join together the *Ka* and the *Ba* while in life and, through a process of spiritual purification, to effect transformation into an *Akh*, or a spirit being. The *Akh* was shown in hieroglyphs as either the phoenix, or as a crested ibis or grey heron, representing the higher intelligence of an individual.[12]

What is significant is that the word *akh* is itself also in the term *Akhemu Urtchu*, meaning the 'never-resting stars', and in the term

Akhemu Seku, the 'imperishable ones', two descriptions of the fixed northern stars which, including the Pole Star, [Temple,1999, p373, Clark, 2000, p125]. It was these stars that were the destiny of the pharaoh in his shamanic ascent. For reasons that will become apparent later, these Egyptian names for stars have within them the all important word *khem*.

Another name for the phoenix, the *bennu* bird, translates as 'resplendent light', which is not that surprising given the connection between the phoenix and the *Akh* [Clark, 2000, p301]. In popular mythology the phoenix is of course a strange bird that cyclically burns itself and comes to life again, and in Egyptian culture the phoenix was associated with long cycles of time, with aeons [Collins, 1998, pp180-1]. According to writer Laurence Gardner, the phoenix was 'burned to ashes in the Temple at Heliopolis, [and] from whose ashes came the great enlightenment' [Gardner, 2004, p28]. It was on the mound at Heliopolis that the Egyptian creation myth states that phoenix was said to have alighted at dawn on the first morning [Collins, 1998, pp235-6, p180].

The root word *bnn* in *bennu* bird is an indication of the fundamental nature of this symbolism, as it has the notion 'to beget' – or 'embryo' in its female form, *bnnt*. It also appears in the Egyptian word for the capstone of the pyramidion, the *benben* stone. The *benben* stone at the ancient Egyptian city of Heliopolis was supposed to mark the site of the primeval mound [Clark, 2000, p68]. The descent of the pharaoh during the *Heb Sed* renewal ritual into the realm of Sokar, deep underground in pyramid, was thus an attempt to recreate the function of the 'primal mound'. There is no doubt that the Egyptians believed that this process occurred at a very fundamental level involving the essence of life, and that the rituals of the *Sed* festival were a re-enactment of what occurred in what the Egyptians called the First Time, the *Zep Tepi*, when the primeval mound came into being.

The Pharaoh Transfigured

This transformation process of the pharaoh into an *Akh*, a spirit being, while still alive, had a serious purpose which needs to be viewed in the context of the pharaoh's responsibility for ensuring that civilization continued, that *ma'at* prevailed. It was while in this state of being, which Rosemary Clark describes as 'an imperishable form of light', that the pharaoh's soul possessed the ability to 'transcend time and space, with the knowledge of all things earthly in the past, present and future'. In this form the soul could communicate with the *Aakhu*, other transfigured beings, and move through cosmic worlds, just like a shaman when he or she is in a trance [Clark, 2000, p125, pp300-01].

In later texts toward the end of the second millennium BC, the title of *The Book of the Dead* apparently began to contain the word *sakhu*. In these texts the *akh*, the 'transfigured spirit that has become one with the light' [Baigent, 1998, pp191-2], could take its place in the *Sahu*, a circle like that on top of the Egyptian cross symbolizing eternal life, the *ankh*. *Sahu* contains the word *sa* which means 'fluid force of the universe' and is both a destination and a state of being. As a destination *Sahu* is the 'dwelling place of all souls', and as a state of being it translates as 'those who are free, noble and accomplished' [Clark, 2000, pp302-03]. The pharaoh was in effect sent on a mission to find non-empirical knowledge essential to the civilization project.

Given the link between stars and transfiguration, it makes sense that the Egyptian word for stars, *khem*, might well be the root of the word 'alchemy', because alchemy, as we shall see, was key to that process [Temple, 1999, p373; Clark, 2000, p125]. But before trying to explain alchemy, first it is important to look in more depth at specific aspects of the *Sed* festival, and in particular the physical location of the festival – a pyramid complex. These are often locations that we have dismissed as burial places but – whether pyramids, ziggurats or simple 'burial' mounds – we shall see they had another function.

Chapter 10

Why Mounds were so Important

Mounds were, without doubt, a characteristic feature of ancient civilization. In one form or another – whether pyramid, ziggurat or earth mound – mounds are to be found wherever there are traces of the ancient model. They were usually the basis of new foundations of the model, sometimes no more than a simple temple on a reed bed surrounded by water. But why were they of such importance to the ancients? Even lesser mounds, such as the *tumuli* that we find in Western Europe, still required a lot of effort to build. Why did they go to such lengths if it were just a matter of burying a king?

Archaeologists sometimes give an unlikely explanation for the building of a ziggurat or a pyramid by saying that the ancients wanted to get closer to their gods. Yet if the ancients had

Pyramid of the Sun, Teotihuacan, Mexico, with mountains in the background

been so keen on the worship of gods from the top of a mountain, they were quite capable of finding suitable mountains. The Sumerians in Mesopotamia did not build a 300ft high ziggurat out of mud bricks in Ur, neither did the predecessors of the Aztecs create the 250ft high Pyramid of the Sun at Teotihuacàn in Mexico because they could not find mountains [Baigent, 1994, p4]. Teotihuacàn is, after all, surrounded by mountains. Taking the Great Pyramid at Giza as another example, its sides were originally smooth, covered in close-fitting marble, which would have made access to the summit difficult anyway. In any case, it is what happened on the *inside* that mattered.

To us a mound or a pyramid is usually a burial site and of little interest beyond what we can learn about a certain culture from its artifacts or skeletons. We have no other way of describing these other than as 'tombs' because we are ignorant of any other function that they may have had. Even though bodies were sometimes buried in them, that did not preclude them from having another purpose.

The long barrow at West Kennett near Avebury in Wiltshire is typical in that it was a relatively large construction, 100m long by 2m high, with the deliberate creation at the western end of a passage with four side-chambers and single end room, opening into crescent-shaped forecourt [Fagan, 2004, p123]. Yet it only ever had 46 people buried in it over a period of 500 years, starting from about 3400 BC. With so few people here, burial was not necessarily its primary purpose.

Bahrain – Another Link with Mesopotamia and Egypt

One of the most impressive examples of so-called burial mounds – if for no other reason than just their extraordinary number – is what has been described as a 'vast necropolis' on the island of Bahrain. Here is an area which has 150,000-250,000 *tumuli* dating from the third or even fourth millennium BC. Some of the 'tombs' are elaborate structures rising to two storeys, 15m (45ft) high.

They are lined with stone blocks two or three deep, with a corridor to the outside [Rohl, 1998, p237, p253, p238]. The less grand 'tombs' are a like a honeycomb network of stone pits crammed together.

What is even more peculiar about this 'necropolis' is that so many of the 'tombs' were never occupied: 'Not a single human bone – or anything else', especially the grander ones, of which nearly 40 percent contained *nothing*. Where bones have been found they date from a later time, from at least the middle of the third millennium BC onward – the Early Dynastic to the Hellenistic periods. Yet the Sumerians and the Ubaid before them, from as far back as the late fourth millennium BC, both left plenty of pottery evidence of their strong presence on the island prior to the mid-third millennium BC. In particular there were temples on Bahrain dedicated to the Sumerian god Enki. It is thought that the tombs were therefore built by visitors from Mesopotamia because Bahrain was their sacred burial ground – called by them 'Isle of the Blessed', *Dilmun* or *Tilmun* – which also was their name for the 'Garden of Eden' [Rohl, 1998, p240, p228 p235, p238].

There are other parallels between Bahrain and Mesopotamia. Archaeologists have been struck by the similarity in construction between a key temple on Bahrain, the temple of Barbar, and the earliest Mesopotamian temple site at Eridu. Excavations down to the earliest level of the temple at Eridu (dating to the beginning of the fifth millennium BC or before) have shown that it was more than just a primitive 3m^2 reed chapel on a pile of earth. What surprised the excavators was to find that those early builders had gone to the trouble of using well-made mud bricks underneath the mound, where they were completely hidden from view [Leick, 2001, p6]. This odd architectural feature of building a reed temple on a mound with a sloping revetment wall of bricks appears also at a third site at Nekhen in Egypt.

Nekhen, as we have already seen, was an important site for

the ancient Egyptians because it was close to the *Per Ur* sacred precinct at Nekheb. Nekhen was known to the Greeks as Hierakonpolis but today is it referred to as *Kom el-Ahmar* or the Red Mound [Rohl, 1998, p349, p359]. This latter name both hints at the function of these brick-built temples and it links back to Bahrain.

Significantly, the reference to red appears in the Greek legend of Erythraeas which, according to archaeologist David Rohl, could have been based on Bahrain. According to this legend, Erythraeas, 'the Red One', was buried 'within a great mound on the island of Tylos' – Tylos being one of the two names for the Bahraini islands and the basis for the name of the Phoenician city of Tyre. David Rohl also makes the point that in the ancient past the Persian Gulf where Bahrain is situated was known by the Greeks as the Erythraean Sea and as the Red Sea [Rohl, 1998, p253]. But what did the Greeks mean by this legend? Could this mention of 'red' being buried in a mound be a reference to some kind of energy source?

A Creative Force

Perhaps part of the answer to these questions lies in some equally strange texts to be found in the Egyptian temple at Edfu. These texts mention that the earliest name of the primal mound was 'Island of the Egg' and that this egg was 'the creative force responsible for the formation of the Earth'. The egg in the Edfu texts was described as being synonymous with the Great Lotus or Throne located within an island, said 'to have emitted a radiance'. A further name for the nucleus of the island was also 'Sound Eye', the 'centre of light which illumined the island'. It was destruction by the Sound Eye of the sacred island that ended the first period of the primeval mound, resulting in it becoming the Place of the Ghosts and Underworld of the Soul [Collins, 1998, p235].

We have seen how the Egyptians regarded a pyramid to be

symbolic of the primal mound; how the *benben* stone had a stepped pyramid as its hieroglyph; and how the phoenix as the *bennu* bird was associated with the fundamental transfiguration of the pharaoh. As already pointed out, both *benben* and *bennu* have at their roots the same word *bnn*, with its idea of 'to beget' (or 'embryo' in its female form, *bnnt*). Central to the whole *Heb Sed* ceremony was the role of the pyramid. The symbol for the *Sed* was itself an image of the pharaoh seated on a dais in the shape of a stepped pyramid [Naydler, 2005, pp90-91]. So, did the Egyptians regard what happened within a pyramid being somehow connected with a 'creative force'?

Interestingly, the Egyptians themselves did not call pyramids 'tombs', but their word was *mer*. This word breaks down into two concepts. The 'm' refers to the idea of place or instrument, and 'r' represented a verb meaning 'to ascend' [Dr Edwards, cited in Baigent 1998, p190] – which of course could refer to the journey of the pharaoh's soul, alive or dead. The word 'instrument', however, hints at a mechanism or process which could have occurred during the *Sed* festival.

Jeremy Naydler is one commentator who questions the conventional belief that pyramids were primarily tombs. He is convinced that their primary purpose was a role in these renewal festivals. First, he points out, they were always built at the beginning of a pharaoh's reign in conjunction with temples and were thus in use throughout the pharaoh's reign. Second, where Egyptian pyramids contain reliefs on the inside, these always refer to the *Sed* festival. Third, the sarcophagi inside them were often empty and, in some cases, even sealed with no evidence of a body ever having been inside [Naydler, 2005, pp117-18]. Where bodies were found in sarcophagi, they were usually either the corpse of a sacrificial bull or a mummy from periods later than the original time of building the pyramid.

Naydler refers to evidence from excavations around the Great Pyramid at Giza that suggested that even all three of these magnificent pyramids were likely to have been involved in *Sed* festivals at the time of their respective pharaohs. There are remains of reliefs and inscriptions relating to *Sed* festival rituals in the adjacent temples and on the causeway from the Great Pyramid. The pyramids on the Giza plateau also have a similar layouts and causeway arrangements linking pyramids to temples as other sites known to have been places for the *Sed* festival [Naydler, 2005, pp103-04].

The Significance of Materials

Before further clarification of the creative force and the *Heb Sed* festival, there are two other key factors to be taken into account in the analysis of mounds – their internal shape and construction materials. The Step Pyramid complex of Djoser at Saqqara in Egypt is a useful starting point because it is known to have been deliberately created as a *Sed* festival site. The Djoser pyramid has passages and chambers underneath it around a 'central granite vault' [Naydler, 2005, pp92-93].

These underground passages and chambers were a common feature of mounds, whether in the Mesopotamian ziggurat or in pyramids in South America. Under the Pyramid of the Sun in Mexico there is a 100m (300ft) long tunnel that leads down to a natural cave under the centre of the pyramid. This cave has been enlarged into a shape like a four-leaf clover to create four chambers of about 20m (60ft) each [Morton & Thomas, 2003, p162]. The Egyptian Great Pyramid at Giza is different because the passage to the chambers ascends rather than descends.

What is more significant are the materials used in the interiors. In Mexico, archaeologists have been mystified by the use of mica hidden *underneath* one of the upper layers of the Sun Pyramid. Although these particular pieces were robbed from the site at the beginning of the 20th century AD, two huge pieces 90ft sq were found intact underneath the stone floor of the nearby Mica Temple. Not only was this an odd use of mica, where it had no decorative function, but the ancients had also gone to a lot of trouble to get this type of mica, which apparently exists only in Brazil, over 2,000 miles away [Morton & Thomas, 2003, p163].

Likewise, in Egypt, the choice of materials for the Great Pyramid at Giza was clearly deliberate – the granite and limestone – as was the construction. The limestone for the pyramid was quarried nearby, but the granite came from Aswan, over 500 miles (800km) away. There are, in addition, a number of design features within this pyramid which do not make sense from a structural point of view, starting with the King's Chamber.

Above the ceiling of the King's Chamber are a hidden series of four layers of granite, one above the other, with gaps between them, and then a limestone gable. Each layer consists of eight or nine granite beams, the largest of which weighs 70 tons, the total tonnage being about 300 tons. These hidden beams have been smoothed on three sides, while the tops of them have been left rough, some with grooves or deep gouges cut out of them.

The granite layers above the ceiling appear to have to *no* structural value and their purpose is not immediately apparent. The limestone gable above the granite beams is supported independently by two limestone walls, not connected to the hidden granite layers. Instead, the granite layers are supported by the granite walls lining the interior of the King's Chamber. These granite walls, in turn, are not connected with the granite floor, which itself has a hidden corrugated surface underneath [Alford, 2003, pp71-80].

The Significance of Shape
The other remarkable characteristic of ancient sacred architecture was the decision sometimes to create curved chambers. It is noticeable that the use of curves occurs in places where hard stone, like granite or mica, were not available, and there was no choice but to use the local materials such as limestone. Some of

the most impressive examples of such a curved shape are the stone-built beehive structures with internal corbelling, known as *tholoi*. These large, manmade constructions could be as much as 8m (25ft) in diameter, with a rectangular antechamber or corridor used as the entrance or *dromos* [Maisels, 1999, p137].

Tholoi are an architectural feature of the early Halaf culture of the highlands of Mesopotamia in the fifth millennium BC. They also appear in Minoan Crete at the beginning of the third millennium BC [Castleden, 1993, p153, p155] and, much later, in mainland Greece toward the end of the 13th century BC. Then they appear in the 4th or 3rd centuries BC where they are to be found in Thrace (Bulgaria), by which time they show the influence of the Mycenaeans, who had much earlier occupied Greece and invaded Crete [Marazov, 1997, pp72-4]. At Mycenae in Greece there are several exquisite examples of *tholoi*, including the Treasury of Atreus, dating to around 1250 BC.

This curved shape is also evident in Malta, in particular at the Hypogeum of Hal Saflieni where there are more than thirty such chambers on three storeys, cut into a limestone hill around a central temple with a corbelled ceiling. The use of this complex, which covers as much as 480m^2, lasted from 3600 to 2500 BC [Rudgley, 1998, pp23-24]. The other prehistoric temples on Malta all reflect enormous curves in their architecture, either apsidal or oval.

Malta - in common with two other areas where rock-cut 'tombs' are to be found, Thrace (Bulgaria) and Phrygia (Turkey) [Marazov, 1997, pp74-75] - has the appearance of megaliths coinciding in the timing of these rock 'tombs'. In Malta the cyclopean building period lasted from 3500 BC to 3000 BC. The largest example of such building is Hagar Qim which has a megalithic temple with a Venus figurine [Rudgley, 1998, pp23-24]. Both Thrace and Phrygia experienced their 'rock-cut tombs' and megalithic phases later, between 1500 and 1000 BC. In Thrace, dolmens, some weighing up to tens of tons, can be found high up on mountain ridges [Marazov, 1997, pp86-87].

Ultrasound and *Djed,* the Divine Sound

So, what were the ancients doing? Why go to the bother to of excavating and transporting difficult and heavy stone like granite from so far away – or mica, for that matter – when these materials did not have any identifiable structural or decorative purposes? What was the point of a curved shape when it would have been just as easy to create a square chamber? Even in relatively simple earth mounds such as the *tumuli* found in Western Europe, writers David Cowan and Anne Silk have noted with surprise the deliberate 'layering of alternate organic and inorganic matter in the roofs of tumuli' [Cowan & Silk, 1999, p74].

Maybe the earlier reference to the destructive force of the 'Sound Eye' is a clue. Several writers suspect that the ancients knew about the uses of ultrasound [Picknett & Prince, 1999, p22; Dunn, 1998; Cowan & Silk, 1999]. Was this the 'creative force' that ended the first period of the primal mound? Commentators Cowan and Silk believe that ultrasound is the reason for the curvature of mounds. They make the point that 'ultrasonic energy… is normally focused by means of either a curved trans-ducer or an acoustic lens. For the former the energy is brought to a focus at its centre of curvature' [Cowan & Silk, 1999, p74]. Properly harnessed, ultrasound is a useful source of energy.

But ultrasound can also be used as a destructive force. As we know from apocryphal stories of opera singers breaking glasses when they sing at a certain high pitch; if something can be made to vibrate at the same frequency as something else, then it can cause it to vibrate and shatter, no matter what its size. Cowan and Silk point to the frequent occurrence of 'acheropita' images throughout the ancient world. These are images of faces that look man-made but they are in fact are a natural phenomenon that happens when two chaotic systems synchronize due to wave

159

energies in a 'boundary condition', such as a well or dish of water, as a result of using ultrasound. Examples of such faces are the Medusa – significant, given her association with horror – and the Green Man [Dunn, 1998, p288].

Nikola Tesla, the Hungarian inventor who in the 1930s did so much to advance our understanding of electricity – he was at one time a collaborator with and then competitor of Thomas Edison – became aware of the destructive force of ultrasound. In one experiment he conducted in New York he created a device that could vibrate at the resonant frequency of a building. But he had to destroy his device quickly when he realized that his experiment was in danger of becoming so successful, by resonating at the same frequency as the steel building he was in, that the building had started to break apart. As a result of his experiment he created panic in the street outside, and had to explain away the damage to the building by blaming an earthquake [Dunn, 1998, pp148-149].

The Egyptians themselves acknowledged the importance of sound to their culture. They had a concept of *djed* that, according to Rosemary Clark, was 'the divine sound which set all forms into motion' [Clark, 2000, p302]. *Djed* was a 'cosmic vibration' animating all matter, bringing it to life. Clark describes 'all living things [as possessing] a distinct frequency derived from the cosmic ambiance which existed at their birth or inception'. The god Thoth, whose name can also be spelt *Djehuti*, was the 'originator of sound or the vibration which set all matter into motion in primeval time' [Clark, 2000, p159].

Clark describes the principle of *djed* as being a 'high science practiced in the temples with chant accompanied by the sacred instruments, the drum and the sistrum'. Knowledge of these frequencies 'enabled the temple mages to invite sacred forces into the sanctuary' [Clark, 2000, p302]. *Djed* was a fundamental resonance, in whatever form that affects living matter – music, chanting or even speech. Like the Bible, ancient Egyptian texts also refer to the importance of 'the Word' in the creation of the world. The first discourse of the *Hermetica* speaks of the 'chaotic dark waters of potentiality' on which Atum's Word fell, calming them and 'making them pregnant with all forms' [Freke & Gandy, 1997, pp38-39].

It is interesting to compare this Egyptian view with experiments carried out in the 18th century when a certain German physicist, Ernst Chaldni, observed what happened to sand on a steel disc. He was fascinated when beautiful patterns appeared in the sand as it was exposed to violin music. Dr Hans Jenny of Zurich continued with Chaldni's experiments after his death, developing something called *cymatics*, the study of wave forms. He discovered, to his surprise, that sound was capable of producing a wide variety of shapes found in nature, such as the honeycomb, snowflakes, starfish and zebra stripes. Specific sounds produced specific shapes, leading him to ask the question 'Are sound vibrations the shaping force of nature?' [Harvey & Cochrane, 1998, p32; West, 1993, p68].

Giza as a Power Plant

Was the Great Pyramid at Giza then a giant power plant that used the *djed* resonance (ultrasound) to generate some sort of electricity? Certainly Christopher Dunn – a craftsman and engineer who has worked with machines for over 35 years, and with ultrasound in particular – writes a plausible account of how the Great Pyramid could have been deliberately built as a giant generator of energy using ultrasound.

He describes the Great Pyramid as acting like any other 'coupled oscillator [which] will draw energy from the source as long as the source continues to vibrate' [Dunn, 1998, p148]. His detailed and technical descriptions have enormous implications for us in our search for renewable energy, not least because the ultimate source of this energy is the Earth itself – taking the ancient Hermetic principle of 'as above, so below' to the extreme.

Dunn points to evidence of an extremely powerful explosion inside the King's Chamber in support of his view that some kind of force was generated inside the Great Pyramid. The traditional explanation that the damage in the King's Chamber was caused by an earthquake does not make sense, as there was no damage to the pyramid lower down. The power of this explosion was so great that it shifted the granite blocks and, he thinks, moved the granite coffer to its current position as well as burning it. He believes that the unusual color of the coffer – a dark chocolate color unlike any granite available in Egypt – and the melted look of the damage to the coffer are attributable to the intensity of the explosion.

Further corroboration that the Great Pyramid could have been an enormous generator of ultrasound comes from closer analysis of the structure of the King's Chamber. Dunn believes that the sole reason for the complexity of the design of the King's Chamber is for acoustical purposes. The empty granite coffer at one end is one potential resonator, but even more important was the specific design detail of *not* connecting the walls with the floor – or the ends of the granite layers above the ceiling with the outer walls. This enabled the whole structure of the King's Chamber to vibrate freely with minimal dampening. The reason for the grooves and gouges in the granite beams above the chamber was the result of 'tuning' the beams to achieve the desired frequency.

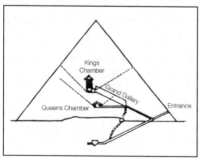

Internal diagram of the Great Pyramid at Giza

The King's Chamber then leads via an antechamber down into the so-called Grand Gallery. The antechamber between the King's Chamber and the Grand Gallery has also been designed specifically with acoustics in mind. It has been constructed mostly of granite, with a number of curious grooves in its walls. The function of this antechamber seems to have been to act as some sort of filter between the King's Chamber and the Gallery.

The Gallery, built on a steep incline over 50m (153ft) long at an angle of 26 degrees has a high corbelled ceiling and no reliefs on the walls, suggesting that it had no ceremonial purpose [Alford, 2003, p107]. It does, however, have a regular series of 27 slots cut into the ramp on each side of the gallery. It is possible that the slots are now empty because whatever they contained may have been made of wood, possibly disappearing at the same time as the King's Chamber exploded – burnt up in the inferno. Christopher Dunn thinks that these mysterious slots could have held a certain kind of resonator known as a Helmholtz Resonator [Dunn, 1998, pp210-11, p165].[13] The grooves in the antechamber therefore could have housed adjustable baffles so that the Helmholtz Resonators in the gallery delivered the correct frequency to the King's Chamber [Dunn, 1998, p173].

Dunn thinks that the ancients were able to tap into the fundamental frequency that occurs as a result of electromagnetic activity between the Earth and the upper atmosphere, known as the Schumann Resonance after the German physicist who identified the phenomenon in the 1950s [Dunn, 1998, p128]. This frequency is very low – thought to be no more than 7.83 Hertz – but it connects with other frequencies higher up the harmonic scale, given the right amplification.

There is a view that the basic frequency of the Earth is in the key of F# and that this is the key to which shamans have traditionally tuned their shamanic drums. In 1996 Tom Danley, an acoustics engineer, carried out tests using sophisticated equipment inside the Great Pyramid at Giza. He discovered that the prime resonant frequency of the King's Chamber at the heart of the pyramid was very low, at around 30 Hz and possibly lower; and that the chamber itself is tuned roughly to an F#

chord [Alford, 2003, p245, p260].

Christopher Dunn bases his opinion upon an assumption that they used the Queen's Chamber below the King's Chamber to create hydrogen gas. The hydrogen gas then rose up through the Grand Gallery to the King's Chamber and became 'excited' by the ultrasound (amplified with the Helmholtz Resonators). The quartz crystal in the granite of the King's Chamber then acted as a transducer, converting the energy into electricity. Dunn then envisioned that the electricity produced in the King's Chamber could be transmitted from there using the kind of wireless technology that Nikola Tesla later worked out in the early 20th century, using sympathetic resonance.

Dunn, as someone who has had personal experience of ultrasonic drilling, has also identified traces of such drilling on artifacts around the Great Pyramid. He believes that ultrasound is the only practical explanation for the ancients' ability to achieve their incredibly precise work on hard stone like granite, diorite or basalt, for large pieces like the granite sarcophagi as well as the smaller, beautifully-worked urn-shaped bowls with narrow necks. They could achieve all of this at a time when the Egyptians were known to have had only copper tools.

Dunn explains that, although the Victorian archaeologist Flinders Petrie found deeper grooves cut in Egyptian quartz than feldspar, he was unable to identify them as being the result of ultrasound drilling because this technology had not yet been discovered in the nineteenth century AD. [Dunn, 1998, p73]. According to Dunn, the ancients would have found it easier to drill through hard granite containing quartz because it 'would be induced to respond and vibrate in sympathy with the high-frequency waves' of the ultrasonic tool bit [Dunn, 1998, p87], more easily than the softer feldspar.

Whether or not Dunn is right about the Great Pyramid, quartz crystal, if cut 'parallel to its electronic axis', will vibrate at a particular frequency that produces electricity [Morton & Thomas,

2003, p54]. Authors David Cowan and Anne Silk confirm that when quartz is cut across its width it can be used for ultrasonic propagation, with lower frequencies being produced [Cowan & Silk, 1999, p74]. Quartz has piezo-electric properties that respond to pulsating rhythms, such as those that occur with ultrasound (*piezo* comes from a Greek word meaning 'to squeeze').

Quartz is used extensively throughout today's electronics industry precisely because of its ability to maintain a constant and accurate frequency. These special properties of quartz explain why it was so important to import the granite containing over 55 percent silicon-quartz crystals from the Aswan quarries 500 miles away. The dressing of the granite rock for use in the King's Chamber would also have helped with 'piezo-tension', increasing the likelihood of cutting the quartz along the necessary axis.

Why the Need for Electricity?

Whether or not there is any truth that the ancients were able to generate electricity on any significant scale, there is concrete evidence that they could at least produce small amounts of electricity. In the 1930s an Austrian archaeologist, Dr Wilhelm König, found in Iraq a 2,000 year old artifact, now called the 'Baghdad Battery' which looks like a small clay jar [Gardner, 2004, p105]. It turned out to be a small electrical device that worked on a galvanic principle using an acid, either citric acid or vinegar, and an iron rod inside a copper cylinder. Reconstructions have been able to produce between 1.5 and 2 volts of electricity. Furthermore, there are items of jewelry which provide proof that the ancients were able to make use of these small batteries in electroplating objects with gold or silver.

Nevertheless, generating electricity on a bigger scale using magnets is not that difficult to do. It was Michael Faraday who established in the 1830s that a changing magnetic field can induce electric currents, simply by moving the magnet near a coil of wire and keeping it moving [Langone, 1989, p75]. On the one hand, electricity creates a magnetic field; on the other hand, magnets can produce electricity.

Magnetite, a naturally occurring mineral, is magnetic and so is meteoric iron. The Egyptians knew about magnets – gold from the Nubian Desert contained magnetite [Hatcher Childress, 2000, p76] – and they had a term for a lodestone that meant 'north-south iron' (*res mehit ba*), emphasizing its magnetic properties. Plutarch, writing in his 1st century AD work, *On Isis and Osiris*, commented that the Egyptians called a lodestone 'Bone of Osiris'. He went on to say that they called iron:

> 'Bone of Typhon', for just as the iron is often, like something alive, attracted to and following after the lodestone, but often turns away and is repelled from it in the opposite direction [Temple, 2000, p283].

'The Egyptians also knew the difference between meteoric iron and smelted iron' [Hatcher Childress, 2000, p76]. Although iron was not widely used in the original model of ancient civilization, it was highly valued. It is thought that the Hittites were among the first to have the secret of smelting iron from about 2700 BC. The Hittites also had something of a monopoly on the trade in iron. Apparently, in 1200 BC the price of iron was 'five times that of gold and forty times that of silver' [Hatcher Childress, 2000, p73, p78]. Alan Alford even talks of the Egyptians having a 'cult' of meteoric iron [Alford, 2003, p168].

Ultrasound may therefore not have been the only technology involved: electromagnetism in some form or other may also have played a part. Certainly, ultrasound explains the shape and style of construction, as well as the choice of materials used – especially the granite containing quartz crystal. It also explains why archaeologists found mica, a good insulator, near the top of pyramids in Mexico, and why earth tumuli were sometimes built with alternate layers of organic and inorganic matter, especially if electromagnetism was involved. Writers Cowan and Silk give the following explanation for such layering. It creates:

...a parallel in physics known as giant magnetoresistance. In this, the resistivity of a layered structure of alternate magnetic and non-magnetic materials dramatically changes when a sufficiently high magnetic field is applied. The effect seems to be due to the fact that the magnetizations in the alternate layers are anti-parallel [Cowan & Silk, 1999, p74].

But apart from powering tools to achieve stonework of incredible precision, or creating gold artifacts to a very high standard, why would the ancients be at all interested in electricity? It is possible that they may have had some kind of electrical lighting. It has often been commented that there is no evidence either of smoke from torches or of brackets to hold up torches in the interiors of tombs for example in the Valley of the Kings in Egypt, enabling people to work on intricate decoration where there was no daylight. Even static electricity can be captured to make a basic device glow to give off sufficient light to see by. But however civilized they were, they were not like us, surrounded as we are by so many electrical appliances: they had no air conditioning, no refrigeration, toasters, electric kettles or irons. So what was the point of electricity beyond a limited use in a minor way?

Chapter 11

The Use of Alchemy in the Shamanic Process

Sacred ritual was, without doubt, a plausible explanation for the ancient use of electricity. If we continue to examine the Egyptian *Heb Sed* festival we can find evidence of the deliberate use of electricity in several aspects of it. In summary, the secret process that the pharaoh underwent in order to make his enquiries as an *Akh* involved his entering an underground chamber in the pyramid. Before that he had participated in a ceremonial procession around the *Heb Sed* festival complex following the standard of the jackal god known as *Wepwawet* (*Upuaut*) or the 'Opener of the Ways', and then consumed a sacred meal, the *hetep*.

Once in the chamber he was given the *qeni* garment to wear and adopted the crown, robe and beard of Osiris. He then lay down either in a sarcophagus or on his special golden *Heb Sed* lion bed (it is possible, for instance, to see Tutankhamun's golden *Heb Sed* bed, found in his tomb, in the Cairo museum). With the pharaoh were normally only two types of priest: the priest who carried the *qeni* garment, and another who had the iron adze of *Wepwawet* for the ritual of 'the opening of the mouth'. Sometimes the *sem* priests who conducted funerary rituals were also there [Naydler, 2005, p72].

During this part of the *Heb Sed* ritual, inside the pyramid, there may have been a desire to create an electromagnetic field around the pharaoh, while he was in trance. The creation of electrical fields possibly explains the *qeni* garment and the *Wepwawet*. Writer Jeremy Naydler describes the *qeni* garment as being the most significant part of the ritual. It involved the priest placing a special garment on the 'living king' over the chest and back, tied up on the shoulders. The *qeni* garment symbolized the embrace of Osiris and emphasized the renewal and reinvigoration of the role of the pharaoh as a Horus-king, thus 'filled with the spiritual potency of Osiris'. It is through this embrace of Osiris that the pharaoh achieved transfiguration as a 'shining spirit' or an *akh*. As the Pyramid Text says,

> O Osiris, this is Horus in your embrace, and he protects you. He has become an *akh* through you, in your identity of the *akhet* from which the Sun emerges. [Naydler, 2005, pp68-71]

While there is obviously great symbolism in the lines quoted above, was there also a more practical interpretation? Was the pharaoh given the *qeni* garment to wear over his chest in order to protect himself, and more particularly his heart, while he was exposed to some kind of electrical or magnetic field in this chamber?

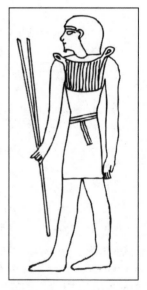

It was after putting on the *qeni* garment that the pharaoh was taken into the secret part of the festival where he lay down in the chamber – in the image of Sokar, the deity of the Underworld – either in a sarcophagus or on a

The *qeni* garment

ceremonial bed with the head and legs of a lion [Naydler, 2005, p71, p76]. Was the reason for using either a granite or alabaster sarcophagus, or a wooden bed covered in gold leaf, the electrical conductivity of these materials, like the granite lining the chamber with its high quartz content?[14] There are Coffin Texts which arouse suspicion that electromagnetism might have been involved.

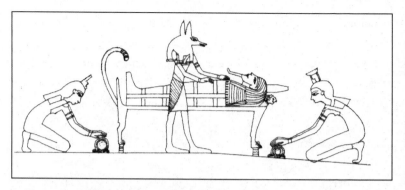

Pharaoh on gold leaf bed being administered to by Anubis, Isis and Nephthys

One text appears to refer to a meteorite being placed inside *Rostau*, the Egyptian name for the pyramids at Giza. It mentions a 'sealed thing… with fire about it, which contains the efflux of Osiris…'; and it goes on to say that 'it has been hidden since it fell from him, and it is what came down from him on to the desert sand. It means that what belonged to him was put in *Rostau*' [Spell 1080 of the *Coffin Texts*, quoted in Alford, 2003, p178].

Another Coffin Text with many references to iron gives an insight into what this might mean. This second Coffin Text mostly concerns the Egyptian ritual of 'opening the mouth' using the iron adze of *Wepwawet*. The text also makes passing reference to the iron 'which Sokar spiritualized' and 'which raises me up, which lifts me up, so that I may open the land of the West in which I dwelt' [Spell 816 of the *Coffin Texts*, quoted in Alford, 2003, p174].

By 'spiritualized', did the text mean *magnetized*? By 'raising up', did it mean the use of magnetized iron in the shamanic experience of the pharaoh? Did this use of magnetized iron involve the *Wepwawet* adze? Have we misunderstood 'opening of the mouth' when in fact it might have meant '*keeping* open the mouth'? If the pharaoh were lying on his *Heb Sed* bed in a shamanic trance, then the iron adze might have been necessary both to conduct a mild electrical current and perhaps to stop him from swallowing his tongue.

Hathor and the Solar Bread

All shamanic journeys begin with some means of falling into a trance. In the case of the pharaoh the sacred *hetep* meal was obviously a key part of that process, *before* he descended into the chamber to undertake his astral-planing. Writer Jeremy Naydler describes this sacred meal as taking place in the Hall of Eating

following the presentation of ritual offerings representing the produce of the land. He calls this an 'offering meal' that 'empowered' the pharaoh and prepared him for the next phase of the *Sed* Festival.

The essential ingredient of this meal, the special 'solar bread', clearly had special properties. One Pyramid Text, which Naydler describes as an 'enigmatic food spell', is an utterance for the offering bread to 'fly up' [Naydler, 2005, p282]. Images of solar bread can be found throughout Egypt – such as a 19th Dynasty relief of the pharaoh offering the solar bread, a conical shaped object, to Anubis the jackal-headed god of the Underworld at the Temple of Abydos. The link with the cow goddess Hathor is, perhaps, of the

Offering solar bread to Horus holding the *Wepwawet*

The solar bread

Hathor as the cow goddess

greatest significance in revealing more about this 'bread'.[15]

A stela was recovered from her temple at Serabit el-Khadimas in the Sinai peninsula showing the pharaoh Tuthmosis III offering a conical loaf to the god Amun-Re, with Hathor present. It had the explanation that he is 'presenting... white bread that he may be given life' [Gardner, 2004, pp15-16].

It was the archaeologist Flinders Petrie who unexpectedly discovered this temple at Serabit el-Khadimas in 1904 while on an expedition to survey ancient turquoise and copper mines in the region. What he stumbled across was a natural cave enlarged into a magnificent manmade site extending over 75m (230ft) in length for sacred purposes. From the inscriptions on the walls, dating back to 4th Dynasty pharaoh Sneferu, about 2600 BC, and other artifacts in the temple, he was able to establish that this temple had been in use for about 1,500 years, only falling into disuse in the 12th century BC [Gardner, 2004, p4]. But what surprised him most was the discovery of 'more than fifty tons of clean, white ash' hidden under flagstones [Osman, 1998, p129]. This ash proved to be mysterious when tested because Petrie could not find evidence of either plant residue or animal remains, indicating that it wasn't ash from animal sacrifices [Gardner, 2004, pp11-12].

Writer Laurence Gardner is convinced that this powder is the same as the biblical description of the *manna* that the ancient Hebrews lived on during the Exodus [Gardner, 2004, p9, p24].[16] For him this powder is the 'flour' that makes the 'bread of life', the 'shewbread', the 'holy bread', the 'bread of the Presence' referred to in the Bible [Gardner, 2004, pp15-16, p23; I *Samuel* 21:6]. He also refers to a New Testament passage in *Revelations*

172

that links *manna* with a white stone: 'To him who conquers I will give some of the hidden manna, and I will give him a white stone...', *Revelation* 2:17.

So, what more can we find out about this strange white powder which may have been the basis for the solar bread, that important ingredient of the *hetep* sacred meal? Electricity may also have been the means by which a key ingredient was prepared for the *hetep*, the sacred meal, in the form of a special 'bread' that the pharaoh consumed. Hathor's better known temple at Denderah in central Egypt could well shed further light on the subject.[17] Hathor as the common factor between the white powder found at Serabit el-Khadimas and the temple at Denderah was not a coincidence.

Part of the significance of the Denderah temple is a large relief of the hieroglyph for gold (a bowl with a necklace draped across it) under Hathor's cow-head face on the rear outside wall. Another name for the Egyptian cow goddess was *Nub-t*, which means 'the gold' or 'the golden' [Clark, 2000, p100]. There is also a curious reference that indicates a link between gold and powder in a letter that the Babylonian king Ashur-uballit I (c.1365-1330 BC) sent to the Egyptian pharaoh Amenophis IV. In it he says, '...Gold is like dust in your hand. One simply gathers it up. Why does it appear so valuable to you?' [Leick, 2001, p207]. Although gold is obviously neither a powder nor white in color, there is nevertheless a connection. Denderah itself provides further clues in the form of some strange reliefs in one of the crypts that look like light giant bulbs. What joins together the Denderah 'light bulbs', gold, powder and the color white, apart from Hathor, is *alchemy*.

The Truth about Alchemy

Alchemy has long been shrouded in mystery. It is usually dismissed because it belongs to a pagan world of magical super-stition and is therefore not for serious study. After all, why would anyone want to waste their life trying to turn a base metal like lead into gold? We have, however, been deliberately misled.

No lesser figure than the father of modern science, Sir Isaac Newton, was privately obsessed with the subject of alchemy in the 17th century, considering it to be worthy of both study and practice. A near contemporary, John Harrison, claimed in his publication *The Library of Isaac Newton* that Newton owned 138 books on 'pure' alchemy by the time of his death [White, 1997, p119] and, when his papers were auctioned in 1936, 121 lots concerned alchemy [Baigent 1998, p214].

Alchemy was also an extremely serious subject for the ancients. In order to understand what it really was and what it meant to the original concept of civilization, we need to consider first what happened in these alchemical processes.

Alchemy has meanings on different levels. It can be defined simply as a 'process of transformation'. On the one hand, 'alchemy' has a spiritual aspect which meant the dissolution of bodies and the separation of the soul from the body; the stripping of matter of its qualities in order to arrive at perfection [Lindsay, 1970, p63]. On the other hand, 'alchemy' is at the root of our word 'chemistry' and contains the notion of creating change through a combination of elements that react together – often using heat to provoke the change. Traditionally, the creation of the Philosopher's Stone, or the Great Work, was the ultimate aim of alchemy.

It was, however, the combination of the metaphysical and mundane aspects of alchemy, which occur at every stage of the process, that made it inaccessible to the uninitiated. Even the descriptions of what are clearly practical processes were discussed in highly symbolic language; deliberately construed in order to protect this ancient art – and the reason for the popular misconception about turning base metals like lead into gold. In truth, alchemy was actually the process by which gold was purified, sometimes using a base metal like lead in that purification process, in order to achieve the Philosopher's Stone. But, as alchemists have been careful not to reveal too much, what can we piece together?

Mouni Sadhu, writing in the late 1950s, gives a good summary in typical alchemical language, albeit repeating the widespread deception that gold results from other metals:

> In order to transform another metal into silver or gold, we must first destroy the imperfect combination of its components, that is to separate the subtle (sulphur – fire) from the gross (mercury – water) in that metal, and then to establish a new, perfect combination, passive or active. The Emerald Tablets speak about this separation of the subtle from the gross. The base sulphur and the base mercury are neutralized by the base-salt. [Sadhu, 1962, pp439-440].

Sadhu goes on to explain in more detail the different stages that an alchemist might follow to create the Philosopher's Stone, which he identifies as being a 'powder'. The first phase is the preparation of the 'universal solvent' or Mercury using a mineral called the 'Magnesia of the Sages'. The second phase is 'the operation' in which at one point produces a 'dazzling white color' after slow heating in the athanor, the alchemical furnace [Sadhu, 1962, pp440-41]. There are several points to note in this account: his view that the Stone is a powder; the specific description of the powder being 'dazzling white'; and that the 'universal solvent' is called the 'Magnesia of the Sages'.

With respect to the Stone as a powder, a 17th century British alchemist, Eirenaeus Philalethes, known to Newton, Boyle and Ashmole, makes clear that the Stone is already gold. It is, he says, 'called a Stone by virtue of its fixed nature; it resists the action of fire as successfully as any stone. In species it is gold, more pure than the purest; it is fixed and incombustible like a stone, but its appearance is that of a very fine powder' [Gardner, 2004, pp14-

15].

Furthermore, there is another consequence to creating this powder. Writer Laurence Gardner claims that it produces weightlessness. He cites a curious medieval tale regarding Alexander the Great and the 'Paradise Stone'. In the story Alexander the Great goes on a journey to paradise where he meets an old man who gives him a special stone. This stone was said to give 'youth to the old' (the Philosopher's Stone has another description as the 'Elixir of Life'); it was also claimed to 'outweigh its own quantity of gold, although even a feather could tip the scales against it!" [von Eschenbach, 1980, p28, p431].

According to Gardner, when the molecular structure of the gold changes, some of its matter turns into energy and becomes pure light – hence the *dazzling* white. Not only does the powder weigh less than the original metal, but it is capable of transferring some loss of weight to the container in which it was originally weighed. He thinks that this levitational effect of the powder is the explanation for the ancients' ability to circumvent gravity when building their massive structures [Gardner, 2004, p130].

In his description of the first phase of the operation, the use of the 'universal solvent', Sadhu gives a more precise definition of what he calls the 'Magnesia of the Sages'. He states that terms such as 'Steel of the Philosophers' or 'Magnesia of the Philosophers' come from 'a mysterious operation involving the use of electricity or personal magnetism' [Sadhu, 1962, p440]. Laurence Gardner explains why electricity might have been used.

There are many objects from the ancient world that are covered in gold leaf – the 4,000 year-old Mesopotamian 'ram caught in the thicket', found in the Royal Tombs at Ur in Mesopotamia and now in the British Museum, is just one example. Gardner states that gold leaf is only possible if all the impurities have been removed first, and there are only certain methods for removing the impurities in gold which the ancients could have known [Gardner, 2004, p118].

One method is cupellation, which uses lead and high temperatures, and another is 'parting'. 'Parting' uses acid, such as sodium chloride, saltpetre, elemental sulphur or antimony sulphide, mainly to remove silver from gold. But the way to achieve completely pure gold, Gardner says, is electrolysis, because only that separates out the platinum group of metals (PGM) from the gold. He says that, when impure gold is exposed to a 'high heat from a DC electric arc', first of all there is a

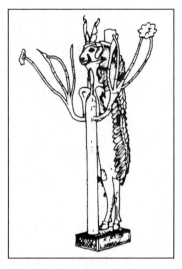

Golden ram in the thicket

flash of light and the final outcome is a bright white powder [Gardner, 2004, p121].

Whether or not Gardner is right, this use of electricity could be relevant to the reliefs in the crypt of the Hathor Temple at Denderah. These reliefs show Egyptians holding what look like giant 'light bulbs' supported by *djed*-pillars. They have every appearance of being some sort of electrical device. Inside the giant bulbs are serpents that represent a glowing. Several writers have commented on the resemblance between these 'light

bulbs' and a modern day Crookes tube. The bulbs themselves are connected by what look like braided wires coming from a box. On the box is a figure seated with open arms and a solar disc on its head – was this figure acting as a 'van de Graaff

Detail from the Hathor temple reliefs

generator', an apparatus which collects static electricity [Dunn, 1998, p232]?

The Hathor temple at Denderah was not, however, the only possible source of ancient electricity. There was one especially famous portable device that may well have been designed to produce electricity. In particular, it could create a powerful enough DC electric arc (a single-directional current between two electrodes) that could kill and could change gold. This device is known to us as the biblical Ark of the Covenant.

The Electrical Ark

It is more than probable that the original Ark of the Bible was an Egyptian artifact; another ark, for example, was found in the

Levite priest with *ephod*

tomb of Tutankhamun, corresponding to the dimensions and construction of the biblical Ark. Whether or not the Tutankhamun ark was capable of the same feats as the biblical Ark is impossible to establish. What we do know from the Bible is that the biblical Ark could produce enough force to kill people.[18] The Levite priests are given special instructions to wash their hands and feet after contact with the Ark, 'Lest they die" [*Exodus* 28, *Exodus* 30:20]. *Exodus* mentions the special clothing needed for the priests involved with the Ark, including a breastplate, the *ephod* – like the pharaoh's *qeni* garment – which all contain lots of gold. In addition, the biblical Ark was able to hover just above the ground [Gardner, 2004, p31].

The design and construction of the Ark is most specific. We are told in the Bible that it was made of acacia wood, overlaid with pure gold 'within and without', with two gold cherubim on top with their wings overshadowing a 'mercy seat' [*Exodus* 25:10-22]. Gardner is convinced that the reason for this design was to create an electrical capacitator. He states that, with enough atmospheric electricity, 'a capacitator the size of the Ark could charge to many thousands of volts' – certainly enough to kill [Gardner, 2004, pp113-14]. The two layers of gold are 'an excellent electrical conductor' with the acacia wood as the insulator, while the cherubim of gold act as the 'outer electrodes'. All the Ark needed was the presence of a magnetic field to set it in motion; and possibly exposure to ultrasound to turn it into a superconductor.

So far, it has only been possible to create superconductivity at low temperatures because it is only then that the electrons slow down sufficiently to form the pairs of the electrons (Cooper's Pairs) that do not repel each other. Could ultrasound be the answer to creating superconductivity at higher temperatures? The reason for thinking this is that superconducting metals send out phonons like 'subatomic noise' sound waves, which overcome the electrons' natural repulsion [Langone, 1989, p20, p24]. So, could ultrasound be the agent that excites the phonons and so creates the condition in which superconducting could then occur at higher temperatures?

Like any other conductor of electricity, superconductors produce their own magnetic field which, if exposed to another magnet, results in levitation because 'unlike poles repel'. Was the biblical Ark able to hover above ground when placed in the Temple in Jerusalem in the Holy of Holies – a 30ft cube lined with gold – because its magnetic field was reacting to the existence of meteoric iron in the *Shetiyyah* rock on which it was situated [Gardner, 2004, p30]?

Alchemy without Gold
In summary, the Egyptian shamanic ritual involved the use of

179

Hathor providing spiritual
nourishment to the pharaoh

alchemically changed gold. The molecular structure of the gold was first of all turned to powder by subjecting it to a powerful electrical force, and then formed into the 'solar bread'; possibly in the Temple of Hathor as it was her role to provide nourishment for the spiritual journey of the pharaoh. It was then delivered to the pharaoh to ingest in his *hetep* sacred meal and help him make his journey to the fixed stars (*al-khemi*). Over time confusion must have arisen between the pharaoh's astral-planing and the special process of preparing gold, with the result that the term 'alchemy' became associated with the process rather than the purpose.

Of course using electrically transformed gold – within an electrically charged atmosphere – was not the only means of inducing a trance. It was just a more sophisticated, 'civilized' version of shamanism compared with what happened in smoked-filled yurts on the Mongolian steppes or in the jungle huts of Peruvian medicine men. Whereas the principle of shamanism has remained the same everywhere over the millennia, the type of stimulant used to induce a trance has varied in different places.

A common hallucinogen has always been the opium poppy. There was poppy growing in Anatolia, which spread to Crete during the Minoan period. Images of the Poppy Head Goddess, including small statuettes, are common in Minoan Create dating from about 1350 BC. Thebes in Egypt was also 'one of the sources of opium' [Castleden, 1993, pp142-43]. North of Iran, along the northern route of the Silk Road, following the Kopet Dagh mountain range through the Merv Oasis and up to Bokhara, there is a series of early Bronze Age oasis forts with evidence of ritual paraphernalia containing residues of ephedra, hemp and poppy. The ephedra was added to keep priests awake and is still used today, especially for asthma attacks, as an adrenaline stimulant called 'ephedrine'[Barber, 1999, p76, p159].

Writer Jack Lindsay makes the point that some traditions of alchemy emphasize the use of an elixir, or a herbal drug, and a strong link with medicine, instead of metals, to achieve immortal life [Lindsay, 1970, p88]. This emphasis appears in the sacred texts of Persian Zoroastrianism, the *Avesta*, and in the Indian *Vedas*, as well as in Chinese alchemy. The Persian tradition involved the preparation of *haoma* and the Indian one of *soma* as their respective elixirs [Barber, 1999, pp162-4]. Lindsay describes these elixirs as having a 'stock epithet: "from whom death flees"' [Lindsay, 1970, p88].

What is striking about the preparation of these elixirs is that the ceremonies involved echo a more familiar, later, Christian Eucharist. One can see how the Christian ritual could arise from attempting to combine the Egyptian 'bread of life' with the 'blood' of the sacrificial crushed Dionysian or Mithraic grape. This interpretation is not surprising given that early Christianity is known for having been influenced by the wine-based Persian cult of Mithras.

What is interesting is to compare pre-Christian Persian rituals with a rather similar Dionysian rite that occurred in ancient Thrace. In Thrace, wine was associated with a cult and myth of killing, dismemberment and crushing to death of the young god, with the crushed grapes representing the sufferings of the deity. The 'dismemberment' aspect of this cult recalls the Egyptian Osirian rites. It is worth noting that the Thracian cults were especially linked to 'the constellation of Orion, the Bull, and autumn'. Here Dionysius was regarded as the stranger who came from outside to inspire others with his cult, turning worlds upside down, with the ivy and grape vine as his characteristic plants. In this case it was the wine that was 'the sacred drink that brings secret knowledge and immortality' [Marazov, 1997, pp40-41].

Modern Shamanism

If we want to know what effect all these various substances might have had on a priest, a king or anyone, we can study the more recent experience of an American PhD student, Jeremy Narby. He traveled to the Peruvian Amazon to study a native community, the *Quirishari*, as fieldwork in 1985 for a PhD in anthropology from Stanford University. While he was there Narby took the risky decision to personally undergo several shamanic experiences. In this case it was *ayahuasca*, a South American vine and the hallucinogen of the native people of the western Amazon, that was the trance-inducing substance which gave him special insight into shamanism.

The first point of interest is that, while in his hallucinating state, Narby encountered a pair of enormous snakes who were able to communicate with him [Narby, 1999, p7]. Although this odd experience unnerved him considerably, he decided to compare his experience with what he found in the literature on ancient myths and legends, and came to the realization that snakes and serpents have always been seen as the source of knowledge (not least in the Garden of Eden) – and may explain the connection between Egyptian gods, *ntr*, and the proto-Indo-European for snake,*netr*.

But what particularly struck Narby was the visual parallel between the image of entwined snakes, such as in the caduceus, and that of the double helix of DNA. Narby then came to the fascinating conclusion that the parallel was more than visual: it was actual. Shamans, he concluded, were communicating at the level of molecular biology, of DNA itself: 'DNA was at the origin of shamanic knowledge'. He was surprised to discover in the technical literature that the shape of the double helix of DNA is 'most often described as a ladder, or a spiral staircase', which compares with the frequent references made worldwide to climbing a ladder during a shamanic trance [Narby, 1999, p108, p63].

Narby was sure that there must be a scientific explanation for the way DNA could transmit visual information. Knowing that DNA emits photons or bio-photons, which are electromagnetic waves, he wondered if shamans were picking up the 'signal', like radio waves, from the DNA. He reasoned that the *ayahuasca* was able to raise the consciousness of the shamans and enable them to become sensitive to these signals, because it has similar molecular properties to the brain hormone serotonin. He also found out through the technical literature that the bio-photons in DNA, in spite of producing only a weak signal, were also highly organized – similar to an 'ultra-weak laser' – which explained the holographic quality of his hallucinations [Narby, 1999, pp127-8].

Narby thus discovered how native people acquired their knowledge of medicinal plants. These plants were often highly toxic and required critical special preparation before use – otherwise these plants would kill you *before* you had chance to experiment with them and find out their healing properties. Even the *ayahuasca* had to be combined with another plant and boiled for hours before it was of use. The native people also knew precisely which plants to choose out of 80,000 Amazonian plant species [Narby, 1999, p40, pp10-11]. The means by which they knew this non-empirical knowledge, he was told and he personally experienced, was through shamanism.

The Divine Rite

Jeremy Narby's experiences may be controversial but they are, nevertheless, an interesting perspective on the part that shamanism could have played in the ability of the ancients to access non-empirical knowledge. It was this ability that enabled them to implement a common theory across continents, allowing Egypt, the most enduring example of the ancient model, to consistently maintain the civilization project for over 3,000 years. With shamanism and alchemy added to the mix, the ancients thus had all the ingredients for being civilized. They had everything they could ever need: they had both an energy source and the means for finding out more information.

Indeed, we are almost totally ignorant of the ancients' abilities. We have to infer their knowledge from the evidence of physical remains. Our reluctance to accept that the ancients used advanced technology is unfortunately conditioned by our absurd notion of historical progress: *how could they have been so advanced in the past when they were primitive and we are modern*? If one cannot attribute the perfect drilling of very hard stone like granite to something like the practical application of ultrasound, then one is left with just unexplained mysteries. What they called 'magic', we call 'science'. It is my contention that there are scientific explanations for much of what they achieved.

What is so extraordinary about all the hints of modern technologies in ancient civilization – the use of ultrasound, electricity, superconductors and magnetism – is that they had such limited application. They were almost entirely used in the context of sacred ritual. There was no attempt to broaden the use of such technologies for the wider benefit of society.

As late as the 3rd or 4th century AD we have confirmation from a certain Zosimus that all alchemical works were for the pharaoh alone. While he noted the importance of the priests in preserving alchemy, he emphasized that they were forbidden by royal decree to reveal its secrets. These secrets were engraved on *stelae* kept in darkness in the temples, written in highly symbolic characters that were not easily understood [Baigent, 1998, p209]. In the ancient model of civilization there was an assumption that only the king or pharaoh, and perhaps certain members of the priesthood, had a need to know these secrets.

In addition, it was assumed that it was the pharaoh's responsibility as the people's earthly representative to undertake the potentially sacrificial role of performing rituals on their behalf. He was the living link between the mundane and the spiritual. Every time a pharaoh participated in a shamanic experience there was an enormous personal risk to him, that Osiris really would 'embrace' him and he would actually die. It was the responsibility of the priesthood to minister to the pharaoh and to make sure that that did not happen. In this sense, it was not so much the 'divine right' of kings but the 'divine rite'.

It is more than probable that the *Heb Sed* festival and its secret rites were practiced in Egypt for thousands of years. The physical remains of the special *Heb Sed* festival courts associated with different pharaohs over the centuries indicate that the pharaoh's involvement in a shamanic process did not alter much over time. Continuous use of the Temple of Hathor at Serabit al-Khadim in Sinai, site of the discovery of the mysterious white powder, is a further pointer to a long tradition of pharaohs participating in the Osirian rites.

There was, however, one highly significant exception to this traditional role of the pharaoh which occurred in the 14th century BC. Even though it was only a temporary change for Egypt, this exception has had profound and far reaching consequences for humanity, especially in terms of shaping our mindsets and encouraging us in the belief that the end justifies the means. I refer to the time that the heretic pharaoh Akhenaten decided that the Osirian rites were obsolete. He was responsible for introducing an entirely different approach to religion by concentrating to the exclusion of all else on the veneration of the *Aten*, the power behind the Sun. Thanks to him, it is more than plausible that humanity first developed religious ideas of monotheism. Whatever the claims of all three Abrahamic faiths (Judaism, Islam and Christianity), there is a case for stating that they can all trace the origins of their theology back to this one pharaoh's religious experiment. It is, after all, Akhenaten's approach to religion that we in the West have inherited – not the pharaohs' secret shamanic tradition. The links between Judaism and the religion of Akhenaten are more than casual, as I shall explain in the next chapter.

Chapter 12

The Origins of a Big Idea – One God

Westerners take monotheism and its origins for granted. Christians might express gratitude to Judaism because of what they regard as the Jews' courageous and stubborn belief in One God in a hostile polytheistic world. From the Garden of Eden onward, the Bible portrays the Jews as heroic torchbearers for the idea of One God – an idea they were brave enough to take to the pagan Egyptians, to show them the error of their ways. Consequently, the Jewish people suffered much persecution until they finally reached the Promised Land. Thanks to trailblazing patriarchs like Abraham and prophets like Moses, the nation of Israel eventually became powerful under kings such as Solomon and David – a power which would not have been possible without monotheism. Without this initial Jewish determination and bravery – so the thinking goes – we could not have become 'civilized' Christians.

But have we been misled? Has the Bible perhaps not given us the whole truth? Is the history of monotheism not quite what it seems? Apart from the Bible itself, what can we learn from other sources?

Far from being torchbearers, there is no firm evidence that the Jews even considered the concept of One God before the time of Moses, around the middle of the second millennium BC. What is more plausible is that, before Moses, the Hebrews had polytheistic beliefs. Plurality of worship is referred to in the Old Testament. Joshua tells the tribes of Israel that their 'fathers lived old beyond the Euphrates, Terah, the father of Abraham and of Nahor, and they served other gods', and that they should 'put away the gods which your fathers served beyond the river and in

Akhenaten and family with the image of the *Aten*

Egypt, and serve the Lord' [*Joshua* 24: 2, 24:14].

Writer Graham Philips refers to the distinct lack of archaeological evidence of monotheistic religion in Canaan prior to the fall of Jericho around 1320 BC [Phillips, 2002, p102], as demonstrated by evidence found at the site of Joshua's capital at Schechem. When excavated by Israeli archaeologists in the late 1980s and early 1990s, this site revealed three distinct levels of occupation. The earlier levels (a pre-Canaanite phase, the Hyksos from 1800 to 1500 BC and a Canaanite phase from 1500-1300 BC) had many deities buried in them, whereas the later Israelite phase, from 1300 until the Assyrian conquest in the 8th century BC, contained 'pagan' evidence of only one small bronze bull.[19]

It seems just as likely that monotheism did not come to Egypt with the Jewish forefathers but was actually an idea that arose in Egypt in the 14th century BC from an Egyptian source. It was probably the 18th Dynasty Egyptian pharaoh Akhenaten who instigated the monotheistic religion that developed into Judeo-Christianity.

Quite early on in his reign Akhenaten imposed on the whole of Egypt the exclusive worship of the *Aten* – the Sun with many hands in the form of rays, holding the ankh, the symbol of eternal life. Because the *Aten* was a god without an image, apart from the symbol of the Sun, Akhenaten decided that there should be no other gods or goddesses worshipped, including Amun, or any other graven images [Phillips, 2002, p190]. He banned the ritual sacrifice of animals. He ended the Osirian rites regarding life after death. He closed the temples at Thebes and, most unpopular at the time, moved his court to a new city that he built at Tell-Armana, named Akhetaten, or 'Horizon of the Aten'. He believed that religious worship should be open to all and not just the élite [Feather, 1999, p85].

Akhenaten did not *invent* belief in the *Aten*, as there were early signs of worshipping the *Aten* from the time of his father Amenhotep III. But Akhenaten did become obsessed with it. As a result, Akhenaten made radical alterations to both Egyptian society and religion. One of the most peculiar aspects was a new style of art in which all human forms, including himself, are given an oddly elongated form, giving rise to the belief in some quarters that he suffered a strange disease. He promoted monogamy, which was unusual for a country in which the pharaoh had several wives, and 'family values' – there are touching images of him with his wife and daughters.

Many of his ideas found their way into Judeo-Christianity unattributed, including some clearly identifiable influences in the *Psalms*. Psalm 104 has direct parallels with a Hymn to Aten found on the tomb of Ayat Tell-Armana [Feather, 1999, p239]. How was it possible for 'pagan' Egyptian beliefs and practices to become so mixed up with the Judeo-Christian tradition?[20]

Abraham and the *Habiru*

The tangled web involving the Jewish forefathers and the Egyptians starts some time after 1628 BC with the invasion of the northern Nile Delta by a group of people called by the Egyptians *Hikau khasut* (a term meaning 'desert princes'). They are known to us by the Greek name *Hyksos* [Phillips, 2002, p74]. They came to Egypt probably in response to environmental pressures, setting up their capital at Avaris in the northern Delta, with their own pharaohs.

Yet again, climatic disaster was a vector for cultural and social change. This time it is thought that the volcanic culprit was Thera (or Santorini as we know it today), the volcanic island in the Aegean north of Crete. The radiocarbon dating of seeds trapped in the volcanic ash at the town of Akrotiri on Thera gives two clusters – one in the 17th century BC and another in the middle of the 16th century BC [Fitton, 2002, pp32-33]. Whether or not it was Thera that caused the climatic changes, the effect was the onset of cold conditions around the world.[21]

The significance of the Hyksos' arrival was twofold. Not only did they invade in order to take advantage of Egypt's superior level of organized agriculture, causing Egypt's Second Intermediate Period to start, but also the Hyksos had semitic tribes living among them, probably Amorites from northern Syria, known as the *Habiru* or *Apiru*. This was the first histori-cally-attested appearance of the Hebrew forefathers. There are also several later references in Egypt to the *Habiru*, on a tomb in the reign of Tuthmosis III which mentions them as prisoners of war, and in a list of foreign captives found at Memphis in the reign of Amenhotep II, which includes 3,600 *Hapiru* [Phillips, 2002, p79].

These two groups, the Hyksos and *Habiru*, entered Egypt from neighboring Canaan where they had been living together for the previous 200 years. Before that they had come from Mesopotamia where the *Habiru* are known to have been in the Hyksos kingdom of Mari, the important trading point on the Euphrates. Clay tablets dating to 1820 BC, excavated in Mari, record the Hyksos king, Zimri-Lim, referring to a people in his land he calls 'the *Habiru*'. Then in about 1800 BC, the Babylonians captured Mari – no doubt because of its attraction as a trading point halfway along the Euphrates – and the Hyksos left together with the *Habiru* for a new life in Canaan. Later, poor harvests there made life difficult and they attacked Egypt [Phillips, 2002, p78, pp74-75].

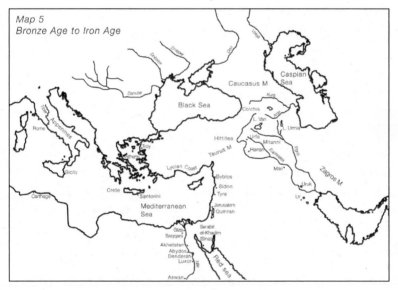

Map 5: Bronze Age to Iron Age

By the middle of the 16th century BC the Egyptians had driven out the Hyksos and regained control of their country. At the end of the 16th century BC, pharaoh Tuthmosis I wanted to prevent a recurrence of the Hyksos invasion, fighting campaigns back in the direction of the Euphrates, crossing the river as far north as the land of the Mitanni where he is supposed to have placed a *stela* announcing his victory. His grandson, Tuthmosis III, consolidated his grandfather's territorial gains by creating an Egyptian empire as far as the Euphrates, including Mitanni in northern Syria. In the end, the Egyptians and the Mitanni established peace treaties and exchanged gifts, with the result that 'three generations of princesses' from Mitanni were married to the pharaoh [Hawkes, 1973, p81].

The special importance of the land of the Mitanni is that this is the most likely place of origin for Abraham and the *Habiru*. There are a number of reasons for believing that they all came from here. In the first place, the name *Habiru* means 'people from the other side of the river', from across the Euphrates [Phillips, 2002, p78]. Mari itself is on the west bank of the river, whereas the land of the Mitanni is on the eastern bank. Secondly, the Bible clearly states that Abraham's father Terah took him and other members of the family from 'Ur of the Chaldeans', and that Terah died in Harran [*Genesis* 11:28-32].

Harran was in the land of the Mitanni, northeast of the Euphrates, and so 'Ur of the Chaldees' was likely to be close by. This Ur has to be a different place from the famous Ur in (cf Golden Crescent Map) southern Mesopotamia excavated by Sir Leonard Woolley, and whose fabulous treasure can be seen in the British Museum. 'Ur', as stated earlier, does not refer to any one particular city: it is an ancient name or epithet with the meaning of 'foundation'.

It makes more sense that 'Ur of the Chaldees' was the same place on the northern Euphrates as the city south of Harran known today as Urfa or Sanliurfa, which has undergone many name-changes over the millennia (it was once called Edessa). The Chaldeans, the ancient group of stargazers, must have been living in Sanliurfa at the time of Abraham. Later, at the end of the second millennium BC, the Chaldeans were known to have been living as group apart, further north in the Commagene, close to Malataya, among the ruins of the collapsed Hittite empire

[Velikovsky, 1978, p176].

The Hittite connection is another reason for believing that 'Ur of the Chaldees' was in the north of Mesopotamia. It was here that Abraham must have had contact with the Hittites. There are several references in the Bible to Abraham's relations with them – for example, he buys a field off them for the burial of his wife Sarah [*Genesis* 23]. If Abraham had been in Ur in the south, how or why would he have had any knowledge of the Hittites? The Hittites were an Indo-European ethnic group which flourished between the mid-17th century BC and 1200 BC, with an empire in the middle of Anatolia, just north of Harran – not in Palestine.

David and Solomon

The relationship between the Hittites and the Hebrews is not the only one that the Bible has been less than clear about. It does not explain the links between Abraham, David and Solomon, and that Solomon, or more likely David, could have been contemporaries of Abraham; nor does it mention that they could have been related by marriage through one of Abraham's descendents, Joseph.

The Bible claims David as the founder of a Jewish royal dynasty, famous for his battles and his empire, and it reveres Solomon for his great temple. But there is no evidence – archaeological or otherwise – that this was ever the case in the 10th century BC or at any other time. There are instead grounds for believing that neither David nor Solomon were 10th century BC monotheistic Jewish kings, but could instead have been the 15th century BC polytheistic Egyptian pharaohs Tuthmosis III and his grandson Amenhotep III. Ahmed Osman, a modern Egyptian writer, has constructed a most convincing case that reveals these alternative identities.

First of all, there are the names. Tuthmosis III's name (Tuthmose) translates as 'son of Thoth'. Osman explains that, in languages without vowels such as Egyptian and Hebrew, 'Thoth' could be written 'TWT' or 'DWD' and would therefore sound like 'David' [Osman, 1998, p21]. In the case of Amenhotep III, 'Solomon' is an epithet with the meaning of 'peace', as in the Arabic word *salaam*: an apt description of a pharaoh who fought few military campaigns and was careful to ensure that the Egyptian empire stayed largely peaceful, by using diplomacy and the exchange of gifts – as evidenced in the correspondence found at Tell el-Armarna in Egypt [Osman, 1998, p68].

Secondly, David's identity is better established through his actions. On the one hand, there is nothing to support the impression that the David and Goliath encounter ever took place. Instead, this story would have been familiar to the Hebrews from their time in Egypt as the Egyptian story of 'The Autobiography of Sinuhe'. Sinuhe was a courtier in the service of Nefru, daughter of Amenemhat I, founder of 12th Dynasty in the 20th century BC. In this story, Sinuhe first flees Egypt, he wanders in Canaan where he has a battle with 'a mighty Canaanite man', he then returns and is buried in Egypt. This story was taught as a literary tale and copies were still in circulation in the 11th century BC [Osman, 1998, pp21-22].

On the other hand, the biblical descriptions of David's battles and his empire match those of Tuthmosis III. Tuthmosis III fought the famous battle we know as Armageddon, which took place at Har Megiddon (Mount of Megiddo) in today's Israel. It was here that Tuthmosis, against all odds, took on the combined forces of Syria and the Canaanites and won. Tuthmosis III stayed in Jerusalem, which became his city – referred to in the Tell el-Armarna letters as *mat Uru-salim*, 'land of Jerusalem'. The Ark carried before him would thus have been the Ark of Amun-Ra and not the one made by Moses [Osman, 1998, p30, pp34].

With regard to the empire, the historical confirmation for God's Covenant with Abraham as reported in *Genesis* has only ever been Egyptian.[22] No other ruler apart from Tuthmosis III established such an empire between the Nile and Euphrates. A commemorative scarab from the time of Amenhotep III, who inherited his grandfather's empire, describes him as a 'mighty king whose southern boundary is as far as Karoy [in Sudan] and north as far as Naharin [the land of Mitanni]'. The decline of Solomon's empire is reflected in Egyptian texts found at Tell el-Armarna, which describe the trouble with the Mitanni in the north and Canaanites in Palestine [Osman, 1998, p61, p72].

If Solomon had been a great Jewish king of the 10th century BC, why would the Bible describe him as being the 'husband of the pharaoh's daughter'? Why would a Jewish king have married the pharaoh's daughter? This description, however, makes sense when one realizes that Solomon was not Jewish but pharaoh Amenhotep III, whose wife Tiye had inherited royal blood through her mother Tuya rather than her father (royal legitimacy in Egypt usually being inherited through the female line), because her father was Joseph.

Indeed, there is some corroboration for the historical figure of Joseph. Ahmed Osman believes that Joseph could have been a vizier of Semitic origin known as Yuya, who ran the royal household at the time of Amenhotep III and his son Akhenaten. The discovery of Yuya's tomb, along with that of his wife Tuya, in the Valley of the Kings in 1905, revealed several unusual features indicating that he was a foreigner. His facial profile was not typically Egyptian and his hands were not crossed in Osirian fashion across his chest, but instead are held in a prayer position under his chin. As someone who was not of royal blood, the explanation for his being buried amongst royalty was that he had held an exalted rank – his wife Tuya was a granddaughter of the pharaoh Tuthmosis III, and their daughter married Amenhotep III, their grandson being pharaoh Akhenaten.

Furthermore, the Bible states that 'Solomon made no slaves' of the people of Israel – which would be odd if Solomon had been Jewish, but not if he were Egyptian [I *Kings* 9:22]. The Bible is careful to point out that it was Jews who administered the system on behalf of the pharaoh: 'The people of Israel... were the soldiers, they were his officials'. The elaborate administrative and taxation system referred to in the Bible was the same as that introduced by Tuthmosis III [Osman, 1998, pp70-71]. In referring to Solomon stationing thousands of chariots and horsemen in the chariot cities, the Bible explicitly states *"And with the king in Jerusalem"* – which again would make no sense if he already was the king in Jerusalem [I *Kings* 10: 26-29].

The Bible further adds to the status of Solomon in its detailed description of the reason for his fame: his building work, in particular his temple and his palace. But the problem for archaeological excavation based on biblical sources is that it has so far been impossible to find these buildings in Jerusalem, partly because of the difficulty of excavation in such a politically sensitive city. While nothing has been found dating from the 10th century BC, there is evidence from the time of Amenhotep III, including a cartouche of his, of construction in Israel at Hazor, Megiddo and Gezer [Osman, 1998, p74].

What is more certain is that the description of Solomon's palace in the Bible matches the great royal complex that Amenhotep III built at Thebes, not in Jerusalem, some time in the 14th century BC. The elements of Solomon's palace – the 'House of the Forest of Lebanon, Hall of Pillars, Hall of the Throne, Hall of Judgment' – as described in the Bible [I *Kings* 7:1-12], were all found here when the site was excavated in 1920. The Festival Hall, built to celebrate Amenhotep III's second *Heb Sed* ritual, corresponds exactly with the biblical 'House of the Forest of Lebanon' with its pillars of cedar wood imported from Lebanon, comparing to the description in the *Koran* of the place where the Queen of Sheba stepped into the pool on her visit to Solomon [Osman, 1998, pp77-8].

Who was Moses, anyway?

Although the Jews have downplayed their Egyptian connections, the Bible does acknowledge that Moses was an Egyptian prince. Whether or not Moses was actually Akhenaten, as some believe, or a son of Akhenaten's, he would still have been a Hebrew descended from Yuya (*aka* Joseph) on his maternal line. *Exodus* describes Moses' mother as being a 'daughter of Levi' [Osman, 1998, p83]. His paternal line would have been Egyptian royalty.

Moses' background was, however, deliberately shrouded in mystery with the 'baby in the bulrushes' story. This was based on an ancient Mesopotamian myth dating to Sargon, as far back as 2800 BC, although it was still well known in the time of the 18th Dynasty. It followed a common theme of the ancient world for foundlings with royal or noble births to have mistaken humble identities and then to be rediscovered in their true light. In the Egyptian version the baby Horus was hidden in a reed boat in marshes to protect him from Seth, his wicked uncle; whereas in the Mesopotamian tale, Sargon tells how his mother 'laid me in a basket of sedge, closed the opening with pitch and lowered me into the river', and he grew up to become a gardener before his true identity was discovered [Feather, 1999, p35].

Moses was also not a real name. In Egyptian, the name *Moses* comes from the word for son – *mos* (as in 'Thut-mos-is'). It would have been understandable for this epithet to have been accorded to the person that the Hebrew tribes deemed to have been the 'rightful son', the one that ought to have been pharaoh. This explanation makes sense in the context of Horemheb, an army general, usurping the pharaoh Akhenaten toward the end of his reign. Horemheb was the pharaoh who then forced Hebrew and Egyptian supporters of Akhenaten into the Delta to work in forced labor under Pa-Rameses, identified in Bible as the persecutor of the Hebrews, who in turn took over from Horemheb after his death at the end of the 14th century BC [Osman, 1998, p197].

Although Akhenaten's son Tutankhamun became pharaoh for a short time, there may have been another son, the rightful heir (the *mos*), who led the Hebrew tribes into the wilderness in the Sinai Peninsula while continuing to believe in the Atenist religion of his father. Here they were able to shelter in relative safety in the temple of Hathor at Sinai, feeding on the unfamiliar white powdered gold (*manna* means 'what is this?'). The golden calf that was turned into more powder could have been an icon of Hathor, who was, after all, the cow goddess. The reason the Hebrews could feel safe in Sinai was because the peninsula was under the governorship of Panahesy, a Levite.

It is something of a surprise to find that Joseph was not the only highly regarded Semite in the house of the pharaoh. We have the impression of Hebrews only being enslaved to the pharaoh – no doubt there were Hebrew tribes in forced labor, building the city of Pa-Rameses and its storehouses. The reality was that, although the Egyptians regained control from the Hyksos, they did not necessarily remove the *Habiru*, who were useful to them, from the running of their country.

The Levite family of Panahesy was one such which achieved elevated status. This family, whose name could be spelt in its Egyptian form as Pa-Nehas and Pinhas, or Pinehas in a Jewish context, held powerful positions within Egypt, in spite (or perhaps because) of its Levite connection. One of them became 'the Royal Messenger in All Foreign Lands and the Royal Chancellor', a prestigious position which included being the non-residential governor of Sinai. Another was Akhenaten's chief priest and chancellor in his new city of Akhetaten [Osman, 1998, p197].

Famous Temples

Individuals like Thuthmosis III, Amenhotep III and Akhenaten (or his descendant) may have been transformed into David, Solomon and Moses. Whoever these characters really were, they are still talked about. There are constant references to Jesus being of the House of David, to the wisdom of Solomon, or to Moses the lawgiver. Whatever ultimately was the truth about the events and people of the 14th century BC, there is no doubt that their impact has resonated through the ages. For some reason, a deep fascination has lasted for a very long time with the Temple of Solomon (*aka* Amenhotep III) – Sir Isaac Newton, for instance, was quite obsessed with every aspect of it.

Yet it was the temple that his son, Akhenaten, built at Tell el-Armana that remained more important for some. Rather strangely, an obscure Jewish sect known to us as the Essenes still retained knowledge of this temple nearly 1,000 years after it was built. Proof has been found at their site of Qumran on the east bank of the Dead Sea among the 80,000 documents attributed to this sect known as the Dead Sea Scrolls. This documentation gave rise to the next shock of the Christian story: the New Testament is no more reliable as an historical document than the Old Testament. Akhenaten, an Egyptian, may have been responsible for the big idea that was monotheism, but it was the Jewish sect the Essenes who were involved in the transformation of monotheism into Christianity – although not in the way that we have traditionally thought.

So why should it matter that the origins of monotheism might not be as taught in modern Western society? It matters because revealing the Egyptian antecedents of Judeo-Christianity highlights the propaganda elements of these religious beliefs, and the need for authority that is so characteristic of the Judeo-Christian tradition. It also matters because the importance that Akhenaten and his Jewish followers attached to monotheism tells us something about their psychological needs, but that doesn't mean that they were correct in their beliefs. It is arguable that if his ideas had not infected the Jewish people as strongly as they did, then maybe monotheism would have been a short-lived aberration. His own experiment in Egypt disappeared quickly after his death and his city of Akhetaten became a pile of stones. The power of his ideas, however, lived on.

However much we might regard monotheism as part of the inexorable progress to modern life, when society is exposed to extreme stress, belief in One God does not prevent people from carrying out appalling acts of barbarity in the name of that One God. In less than 150 years after the Jews left Egypt, they were willingly conducting unspeakable forms of human sacrifice to appease a One God, as the Bronze Age order was taken over by Iron Age chaos.

As we shall see in the next chapter, it was from the time of the shift from Bronze Age to Iron Age onwards that the civilization project became vulnerable. From about 1200 BC onward – around the time of the fall of Troy – life in the eastern Mediterranean began to change. At this time there were tribes on the move who did not fully understand the holistic theory of civilization and who did not know that alchemy was the basis of the relationship between priest and king. These tribes were some of the most influential of our Indo-European ancestors, the antecedents of the Greeks and the Romans. Interestingly enough, it was possibly another natural calamity that brought them into the Mediterranean arena and, to use a well worn cliché, 'changed the course of history'.

Natural disasters have been part of the human experience ever since there have been human beings. What has changed over the millennia has been the way in which people have responded to such disasters. The event that occurred in 3195 BC was probably unusual if indeed it was a comet, with such devastating effects. More typically volcanic eruptions have had adverse effects on populations, causing harvests to fail and people to starve. Given our own concerns today about environmental disaster and our fears for civilization, it is interesting to observe the variations in response that occurred in the eastern Mediterranean and beyond, during the second and third millennia BC.

Chapter 13

Iron Age Strife: the end of Ancient Egypt

An essential point about the original concept of civilization is that the principles, if followed properly, work in bad times as well as good. Ancient Egypt is a good example of the principles in practice. When the Nile floods failed in 2184 BC, ushering in a prolonged period of drought lasting 300 years, the pharaohs lost credibility and there was chaos. Not surprisingly, Egyptologists refer to this time of disruption as the First Intermediate Period. It was nearly 200 years before Pharaoh Mentuhotep I was able to restore order in 2046 BC [Fagan, 2004, p144]. What solved the crisis for Egypt was Mentuhotep's decision to invest heavily in agriculture and centralized storage. The Mesopotamians followed the same strategy and also survived. As Brian Fagan says, 'The strategy of centralization, of an organized landscape, was the best defense against an unforgiving world' [Fagan, 2004, pp144-5].

But even though investment in the management of resources can become critical for survival, difficulties arise for the well-prepared because they are still targets for attack from the less-prepared. When times are tough, people move. They look for better places to live, for a new homeland. What we can learn today from ancient disasters is that we could be faced with equally significant socio-political and cultural change. In the ancient past whole ethnic groups displaced themselves, whether it was Jews into Canaan, then Egypt and then back into Canaan/Israel, Armenians into Armenia, or Greeks into Greece.

Without doubt, Egypt was vulnerable in the 17th century BC at the time of the Second Intermediate Period when the Hyksos and the *Habiru* decided to invade. While one environmental crisis brought the Semites and the Hyksos into Egypt, another one probably forced them out toward the end of the 14th century BC, forming the background to the Jewish Exodus.

Thera could have been the culprit for both the volcanic eruptions of the 17th and the 14th centuries BC [Baillie,1995, p74]. Prof Schoch is of the opinion that Thera did not explode in 'one massive explosion like Krakatoa or Mount St Helens [but] collapsed in slow stages'. People in the town of Akrotiri on Thera had time to escape as no bodies have been found under the ash [Schoch, 1999, p90]. So Thera may have released enough volcanic ash into the atmosphere in the 17th century BC to affect the weather without actually dramatically blowing up, and then it blew up later in the 14th century BC. This may have been such a 'low-sulphur eruption' that it was not detectable in the Greenland ice-sheet, where traces of global-scale disasters are often found [Baillie, 1995, p112].

Writer Graham Philips is convinced that volcanic disruption was responsible for the experiences of the Jews at the time of the Exodus. He compares the biblical story with known natural phenomena that occur as a result of volcanic explosions, including the biblical parting of the waves. Philips refers to a 1956 expedition of geologists from Columbia University who recovered samples of pumice from the seabed which carried over to Egypt on the wind. They also determined that the power of the Thera explosion was greater than that of the atomic bomb at Nagasaki in 1945 [Phillips, 2002, p15, p17]. He attributes the parting of the waves to the withdrawal of the sea before the arrival of a tsunami, which is what could have happened on the Egyptian Mediterranean coast if Thera had blown up then – indeed, it was this phenomenon that killed so many in the tsunami of December 2004, lured onto empty beaches before the wave hit.

However bad the effect was on Egypt, certainly the country was able to recover this time without loss of control. Indeed,

Egypt went on to become even more powerful, as demonstrated in the grandiose building works of Rameses II, such as the magnificent temple at Abu Simbel, and his domination of most of the 13th century BC. The Egyptians were then better able than most to survive the next disaster when the Hekla volcano in Iceland erupted in 1159 BC [Fagan, 2004, p193]. But this time, in the 12th century BC, it was different. This catastrophe, which marked the transition from Bronze Age to Iron Age, as symbolized by the fall of the legendary city of Troy, unleashed a social change combined with a change in ethos that eventually led to the collapse of civilization in Egypt.

Bronze Age becomes Iron Age

This 12th century BC crisis was particularly bad mainly because the impact of the Hekla volcano went on for such a long time. The effects were global. Baillie mentions that tree growth was affected for nearly 20 years, from 1159-1141 BC, showing narrow growth-rings at this time. He states that there is evidence of frost rings in 1132 BC in a foxtail pine chronology discovered in the Sierra Nevada.

The Shang Dynasty in China, having come to power thanks to the previous event in the 17th century BC, now lost power. Reports of catastrophes in Ireland lasted until 1031 BC. In Britain there was widespread abandonment of upland habitations and increased construction of defensive sites. Greece began a 400-year 'dark age' with a prolonged period of poverty and depopulation [Baillie, 1995, p82-3, p158, p150, p82-3].

The situation in the eastern Mediterranean and the Middle East was just as bad, with severe climatic changes occurring in the Mediterranean. A serious decline in rainfall created a problem for the irrigation of crops and water for cities, especially for Phoenician ports built on islands, reliant on water from the mainland. Loss of woodland encouraged a change to a Sahara-type environment in Syria and Palestine [Aubet, 1993, pp54-55].

The process of deforestation had been accelerated by the Phoenicians' willingness to supply their Cedars of Lebanon to whoever would pay – not least to Solomon for his famous Temple. It was, however, social reactions to the climatic conditions that were most appalling.

What few modern analysts seem to have noticed – possibly because of today's philosophy of history that assumes progress – was that this shift from Bronze Age to Iron Age was a step *backward*, both technologically and culturally. Iron tools made from smelted iron ore are easier to make than bronze tools from a complicated alloy of copper and tin. Some blame the Hittites for losing the secret of smelting iron [Barber, 1999, p34] and its widespread use in weaponry. The earliest metals worked as long ago as c.8000 BC had been delicate copper and gold rather than iron, and the first alloys were of copper and arsenic [Mohen & Elvere, 2000, p28]. Bronze had been used in the eastern Mediterranean and northern Europe from 3000 BC until about 1500 BC for prestige objects, arms and jewelry.

The start of the Iron Age heralded a transition to a more brutal time in more ways than one. This time those who went on the rampage in search of food were better armed – bronze swords were no match for iron weapons. The social response to the effects of this volcanic eruption marked the beginning of the decline of civilization. This particular crisis probably lay behind the terror of the mysterious 'Sea Peoples' in the eastern Mediterranean, drawing certain Indo-European tribes – especially Greek-speaking ones – further into this arena.

In Prof Schoch's view, every significant community in the eastern Mediterranean was affected. There were 'big black destruction layers' and widespread disruption everywhere [Barber, 1999, p185]. It is around this time that Troy became the stuff of legends – until Heinrich Schliemann decided in the 19th century AD that Homer had been right, and began with his excavation in 1870 in western Turkey to find the ruins of the legendary city [Mohen & Elvere, 2000, p20]. On Cyprus the three principal cities were burnt, possibly twice; Ugarit was burnt and never rebuilt; no palaces of the Mycenae survived, instead becoming part of Homer's legends; the Hittite empire collapsed. Egypt escaped the flames but was attacked by armed refugees and was weakened, thus ending the New Kingdom [Schoch, 1999, p129].

The response of people was also different, in comparison with the fourth millennium BC cataclysm. Whereas 2,000 years earlier there had been significant population dislocation and much hardship before civilization reasserted itself, this time people regressed into a more primitive and superstitious worldview rather than relocating. One of the most shocking aspects of the reaction to the 12th century BC crisis was that for the first time there is evidence on a significant scale of human sacrifice. This involved an especially repugnant sacrifice involving children, called the *mlk*, *molk* or *molloch*, that went on for a long time among people of the eastern Mediterranean based in Canaan – that is, among Phoenicians and Jews, people who had earlier been civilized.

There are many references to this grim aspect of human nature in the Bible. It was called the *molk* sacrifice because it involved the burning of newborn babies, often the first born, or young children (the word 'milk' being explicit), to appease the god Baal ('Baal' means 'lord' rather than 'god'). After they had been cremated, their ashes were buried in urns in special cemeteries called *tophet* – the Jerusalem *tophet* was, for example, near the city in the valley of Ben Hinnom [Aubet, 1993, p208]. There was a legal prohibition of the *molk* sacrifice in the *Pentateuch* [Aubet, 1993, p210], and one of the reforms of Josiah was to 'Defile Topeth, which is in the valley of the sons of Hinnom, that no one might burn his son or his daughter as an offering to Molech' [II *Kings* 23: 10-11]. What is surprising is that the Phoenicians carried on with the *molk* for such a long time.

One of the consequences of the cataclysm of the 12th century BC was that the Phoenicians began to look for new maritime bases from which to run their trading operations within the Mediterranean and further afield, taking the *molk* with them. Thus, from about 1050 BC, a diaspora from the Phoenician city of Tyre began with the result that, quite soon after, a new base was constructed at Cadiz – or 'Gades', being the plural of a Phoenician word *gdr* meaning 'fortified citadel'. This was an ideal harbor for the Phoenicians, being a series of islands attached to the mainland by causeways [Aubet, 1993, p221, p21]. Cadiz also had the advantage of lying beyond the Straits of Gibraltar, opening up the Atlantic coast to them.

Within 200 years or so, the Phoenicians also had bases on Sardinia and at Carthage in North Africa. These sites were chosen because the predominant currents and wind directions in the Mediterranean made them easier anchorages between Cadiz and the Mediterranean east coast. It is at Carthage that there is especially gruesome evidence of the *molk*. The *tophet* in Salammbo district of Carthage was begun in 700 BC, then remained in uninterrupted use for 600 years [Aubet, 1993, p208]. Even after the Romans banned the practice in 146 BC it continued in secret, and as a result the *tophet* there contains the remains of more than 20,000 cremation urns in nine levels, covering an area of 6,000m^2 [Aubet, 1993, pp211-12]. So, why were the reactions after the 12th century BC volcanic eruption so bad that people agreed to the sacrificing of their most precious firstborn children?

The Sea People

To some extent, the distress of the Phoenicians, and their desire to placate a god through sacrifice, is understandable because they were among the most affected by the invasions of the Sea People. There is no consensus on who the Sea People were, but they are credited with much of the destruction along the east coast of the Mediterranean. One view is that they were displaced people from the Asian steppes, marauders who came west.

But the Hittites in Anatolia, central Turkey, lost their kingdom to the Phrygians, an Indo-European group who had been living further to the west, not the east. It was these Indo-Europeans who became famous for having kings all with the name Midas, and for their capital Gordion where Alexander the Great is said to have cut the legendary Gordion Knot. Likewise, another Indo-European tribe, the Armenians, with close linguistic links to the Phrygians, Thracians and Greeks [Mallory, 1991, p34], also went east and came to occupy the territory of the Urartians. As both Armenia and central Anatolia were landlocked, it is unlikely that either the Armenians or the Phrygians were the Sea People. It could, however, have been another Indo-European group, a predominantly Greek-speaking people known to be seafarers and pirates with the name of Sherden or Sardan.

It is generally assumed that the Sea People are those people depicted in reliefs at the temple of Medinet Habu in Egypt of the pharaoh winning a great sea battle. These tribes are thought to have come from a variety of places including Libya and the Lydian coast of Turkey [Buttery, 1974]. Yet Velikovsky, the renowned international scholar, convincingly argues that it is not possible for the Medinet Habu reliefs to represent battles with the Sea People in the 12th century BC because, he claims, the Medinet Habu temple did not exist until much later.

Even though the Egyptian Medinet Habu scenes may show a much later battle, it is still plausible that one of the tribes depicted, the Greek-speaking Sherden, could have been the original Sea People. The Sherden are sometimes thought to have come from Sardinia. But that seems unlikely given that Sardinia is some distance from the eastern Mediterranean and is not known for playing any significant part in Middle Eastern history. It is more likely that they were from Lydia in Asia Minor. Velikovsky describes these Sherden as being mercenaries from Lydia and that the Lydians in the army of Xerxes had helmets with horns and a crest [Velikovsky, 1977, p54, p56] – much like Vikings with their horned helmets and fair complexion, long hanging moustaches and large earrings.[23]

The Sherden might have been in the area long before the catastrophe of the 12th century BC because, in the 13th century BC, Rameses II had used the Sherden as mercenaries and for his personal bodyguard. There is a curious connection between Rameses II and the Greeks, involving the place known as Colchis at the western end of the Caucasus (now part of Georgia). Greek speakers may have been living in this area of the Caucasus having left the Pontic-Caspian to the north after the fourth millennium BC disaster. Following their encounter with the Egyptians at the southern foot of the Caucasus on the Black Sea coast, it could have been Rameses' continued use of this particular tribe that then drew them further into the eastern Mediterranean arena as the Sea People. But why Colchis?

Caucasian Plant Connections

Colchis is of course famous as the legendary land of the Greek myth of Jason and the Argonauts. But Colchis was also known for being the place where the colchicum plant, known as meadow saffron, grew. The meadow saffron is similar to *crocus sativus*, used for actual saffron, and the yellow dye these plants produce obviously equates with the color of the Golden Fleece. But the importance of meadow saffron to Egypt and elsewhere in the Mediterranean was as a cure for gout [Temple, 1999, p213, p228].

The ancient historian Herodotus believed that the Colchians were of Egyptian descent – he describes them as having 'black skins and woolly hair' – as a result of Rameses II's army having made a base there in the 13th century BC [Temple, 1999, p209]. The Egyptian link with Colchis may even have pre-dated Rameses II as the Minoans were probably involved in the trade bringing meadow saffron to Egypt, using the port at Troy as a safe harbor *en route* to the Black Sea. There is certainly evidence of black and grey Minoan pottery found at an excavation level of Troy relating to the middle Bronze Age [McCarty, 2004].

Further indication of the crocus connection and the Minoans is to be found in the Minoan town of Akrotiri on Thera. Akrotiri was well-preserved by the volcanic eruption on Thera and in the town there is a building described as possibly having a religious function, with a wall-painting of women gathering crocus flowers around a central goddess figure and 'male figures engaged in what is perhaps the ritual presentation of a garment' (the Golden Fleece?) [Fitton, 2002, p168].

While the Greek *Sherden* may have been the Sea People, active on the Lebanese coast, and the Mycenaean Greeks, under the leadership of Agamemnon, may have been responsible – together with their allies – for the collapse of Troy, it was another particularly fierce Greek tribe from the northwest, the Dorians, who are generally blamed for ending the Bronze Age. The Dorians are thought to have wiped out the remaining vestiges of Minoan culture on Crete in about 1200 or 1150 BC, when many sites were burnt and abandoned [Castleden, 1993, p36]. They also ended the reign of the Greek Mycenaeans and invaded that part of today's Bulgaria known as Thrace.

Curiously, this tribe had another Caucasian plant connection in common with the Egyptians – although this time the connection was not contemporaneous but thousands of years apart. Both were influenced in their use of fluted columns by the Giant Hogweed that grows in the Caucasus. Indeed, the Dorians gave their name to one of the five architectural orders of classical design, the Doric Order, based on the fluted column – a style that had already been in use by the Egyptians in the distant past.

The End of Egypt Foretold
The shift from Bronze Age to Iron Age began a period of increasingly difficult times for Egypt and serious threats to civilization. In many respects what happened in the 12th century BC – the

cultural and socio-political changes following a natural crisis – triggered a chain of events that eventually resulted in the final collapse of Egypt. Although Egypt survived the immediate effects of the 12th century BC, the country was undoubtedly weakened.

The megalomania of pharaohs like Rameses II in the previous centuries may have contributed to the gradual decline. Certainly in the centuries that followed, particularly in the 10th and 9th centuries BC, there is a notable degeneration in the Egyptian priesthood due to an over-reliance on a superstitious use of magic, with the reciting of magical formulae and with importance attached to amulets and so on [Feather, 1999, p58]. Egypt was becoming vulnerable again to invasion by others, especially Indo-Europeans, starting with the Persians. There are some particularly sad passages in the *Hermetica* which claim to foretell the disappearance of Egypt as a sacred land:

If truth were told, our land is the temple of the whole world. A time will come when... Egypt will be abandoned. The land that was the seat of reverence will be widowed by the powers and left destitute of their presence. When foreigners occupy the land and territory, not only will reverence fall into neglect, but a prohibition... will be enacted against reverence, fidelity and divine worship. Then this most holy land, seat of shrines and temples, will be filled completely with tombs and corpses... only stories will survive and they will be incredible to your children. Only words cut in stone will survive to tell your faithful works. Whoever survives will be recognized as Egyptian only by his language; in his actions he will seem a foreigner [Copenhaver, 1992, p81].

No doubt one reason why Egypt survived for as long as it did was that the principles of civilization, so impressive to foreigners, were so ingrained in the country after many millennia. Also the temple system that guarded so much of the knowledge was deeply impenetrable. Even when Cambyses destroyed the temples in the latter half of the 6th century BC, Egypt could cope with the Persian invasion because the rituals were safeguarded in the desert oases. Life might have been difficult as part of the Persian Achaemenid empire – and as the epitaph of Petosiris, a priest in the temple of Thoth at Hermopolis, complained, no work was done in the temples because of the foreigners [Velikovsky, 1977, p161] – but the Egyptians struggled on.

Greek Efforts

Apart from trade, the use of mercenaries and times of temporary invasion, Egypt had effectively been a closed country for thousands of years. It was therefore something of a culture shock when pharaoh Amasis deliberately decided to allow the Greeks to develop a trading base at Naucratis in the Delta in the middle of the first millennium BC, and Egyptian temples began to accept Greeks as students from this time onward.

We persist in believing that the Greeks were responsible for so many inventions with regard to medicine, mathematics or architecture. Greek classical architecture in the form that we know it, using stone – the petrification process – only began in the middle of the first millennium BC. Masterpieces like the Acropolis in Athens were built in the 5th century BC *after* the Greeks had learnt from the Egyptians, who had been building in stone for thousands of years. Yet, just because the Greeks adapted what they learned to their own tastes, it is erroneous to assume that it all started with them.

It is thought that Pythagoras – the son of a Phoenician trader born in the middle of the 6th century BC on the island of Samos, with which the Egyptians had been trading since the 7th century BC – spent 25 years as a student in the Temple of the Sun at Heliopolis [Clark, 2000, p98]. Plutarch, the Roman historian writing in the 1st century AD, was of the opinion that Solon, the great Greek lawmaker, had spent time with priests in Heliopolis and Saïs around 590 BC [Dunbavin, 1995, p13]. Demokritos, credited with a theory of atoms in the 5th century BC, was considered by Diogenes, writing in the 3rd century AD, as reputed to have 'learned geometry from the priests of Egypt' and, according to 1st century BC historian Diodorus Siculus, Demokritos 'spent five years [in Egypt] and was instructed in many matters relating to astrology' [Lindsay, 1970, p93].

It was better for Egypt, to some extent, after Alexander the Great invaded in 333 BC. Alexander the Great was so much in awe of the Egyptians that he went out of his way to encourage the building of a new city, Alexandria, which became a great centre of learning, thus saving as much as possible from the Egyptian past, as well as supplementing Egyptian knowledge with requests for manuscripts from the libraries of kings elsewhere in the Middle East [Baigent, 1998, p194]. The great library at Alexandria that his general Ptolemy created became a focal point and a melting pot of different influences from Persian Zoroastrianism and Judaism, to Hinduism and Buddhism from faraway India. The cosmopolitan flavor of Alexandria has lasted for thousands of years with many Greek descendants continuing to live there even today.

Ptolemy and the dynasty that followed him, far from wanting to destroy Egyptian culture, were keen to be accepted as Egyptians, not as Greeks. Much of what remains of Egyptian temples today dates from the time of the Greek pharaohs and, although clearly different, is still in the Egyptian tradition. They continued with the Egyptian religion, even if in a syncretic form that combined Osiris with Apis the bull to create Serapis [Baigent & Leigh, 1997, p17]. The Macedonian Greeks did not seek to impose Greek culture on Egypt, nor did they export Egyptian style wholesale back to Macedonia. For at least 300 years the Greek Ptolemies succeeded in preserving much of Egyptian knowledge and culture, protecting the country from others. It was thanks to this desire to combine Greek and Egyptian, without the loss of either, that the Rosetta Stone was created – our critical code-breaker.

The End

It could be said that the prophecy in the *Hermetica* was not fulfilled for another 1,500 years after the fall of Troy; not until the 4th century AD, when the last surviving Egyptian priests were massacred at the Temple of Isis on the island of Philae, south of Aswan. But in spite of the ferocity of the early Greeks, it wasn't the Greeks, the Sea People or the Persians who destroyed ancient Egypt. It was another Indo-European people, a previously unknown people about whose early existence little is known – even though they claimed descent from a survivor of Troy. They went on to become a much greater power than the Greeks ever had been. These were, of course, the Romans.

The Romans' lack of respect for Egypt was in distinct contrast to the Greeks. In 30 BC Cleopatra was unable to prevent the Romans from taking over. Within a few hundred years Egypt was totally destroyed and the sad prophecy had come true. The last known hieroglyphic inscription was engraved 'on the island of Philae in AD 394' [Fowden, 1993, p64], and 'command of ancient Egyptian hieroglyphics was a thing of the past by AD 400' [Merkel & Debus, 1988, p31]. Not until the discovery and trans-lation of the Rosetta Stone in the 19th century, nearly 1,500 years later, were the wisdom and knowledge of the Egyptians once again accessible to us. Until then any chance of finding out about the true civilization, about living in *ma'at*, in harmony, disap-peared for a very long time. The *Hermetica*'s prophecy continued with these words:

> [People] will not cherish this entire world, a work of god beyond compare... They will prefer shadows to light, and they will find death more expedient than life. No one will

look up to heaven. The reverent will be thought mad, the irreverent wise, the lunatic will be thought brave, and the scoundrel will be taken for a decent person. Soul and all teachings about soul ...will be considered not simply laughable but even illusory...' [Copenhaver, 1992, p82]

What replaced *ma'at* was a new Roman ethos based on power and glory rather than harmony. More worrying still, it eventually became the powerful ideology of a relatively new religion. Significantly, it wasn't environmental problems that destroyed Egypt, but a change in attitude, in ethos. From our perspective, Egypt's ability to survive so long and to weather so many setbacks demonstrates the robustness of the original concept of civilization. It shows that the concept was about more than grain storage and sewage systems. What is so striking about ancient Egypt is that the principles of civilization which had remained recognizable there for more than 3,000 years only really succumbed to the Romans in the 4th century AD.

It is worth wondering, therefore, how an obscure Italic-speaking people, with possible origins on the Pontic Caspian sometime before the fourth millennium BC, managed to become so powerful that it could finally end one of the greatest examples of the ancient civilization model that the world has ever seen. Certainly, by the end of the 4th century AD the Romans had created the biggest empire in the ancient world, with territory stretching from Britain in the west to the Euphrates in the east, and frontiers as far as Germany in the north and most of North Africa in the south – a much bigger empire than the Egyptian empire of the 18th Dynasty. How did they do it?

Chapter 14

Roman Supremacy

We do not know much about the origins of the Romans. Apart from being able to reconstruct their language to proto-Indo-European in the fourth millennium BC, they seem to have left little trace of their journey from the Pontic-Caspian – unlike the Celts or the Greeks. Part of the Roman foundation myth was that they left with Aeneas as survivors from the fall of Troy – but in fact no one really knows how the Romans came to the western side of the Italian peninsula, and there is a 400-500 year gap between the fall of Troy and the rise of Rome.

Both Celts and Greeks can be tracked through toponyms and other evidence. The Celts, we know, entered Italy from the north across the Alps. The earliest archaeological evidence for Celts in Italy is an 8th century BC boundary marker with Etruscan letters showing a Celtic warrior – although they were possibly in Italy earlier, as a result of overpopulation in Gaul. They became known as the Cisalpine Gauls and had territory in the Po valley. They are even considered to have founded Italian cities such as Milan, Turin, Bergamo, Como, Modena, Bologna and Ancona [Berresford Ellis, 1990, p23, p25]. The Greeks had also been in Italy since the 8th century BC, having had a base on the island of Ischia since 770 BC which they had visited in search of metals. They eventually came to dominate southern Italy, which became known as *Magna Graecia* (Greater Greece) – home to Pythagoras and his school.

With no written records, it is impossible to say exactly how or when the Romans arrived in Rome, when Rome became characteristically 'Roman'. But it is certain that the Romans did not arrive on the Tiber with a fully formed model of civilization – in contrast with the Egyptians, the Minoans, the Phoenicians and others, who all seemed to have been 'civilized' from the start. It is more likely that they learnt about civilization from the Etruscans who were already there. The Romans had only a language which contained certain concepts relating to the civilization model; other than that, they had little more than a militaristic approach to life.

It seems probable that the Italic speakers arrived on the eastern side of the Italian peninsula as part of a general Indo-European tribal shift to the west from at least the 12th century BC onward – what Mallory refers to as 'Balkan currents'. The linguistic evidence also confirms that non-Indo-European languages such as Etruscan survived for much longer in the west of the Italian peninsula, implying that the intruders into Italy came from the east.

There is in the archaeological record a marked change from Bronze Age to Iron Age at this time – the rapid spread of the proto-Villanovan culture about 1100-900 BC [Mallory, 1991, p92]. There were significant changes in burial techniques, metal-working and improvements in agriculture – all of which indicate an influx of new people with new ideas [Cornell, 1995, p31, p33].

Rome's Etruscan Origins

The site of Rome itself had been occupied from 1000 BC and there are traces of primitive wattle and daub huts with thatched roofs supported on timber posts from the middle of the 8th century BC. Major developments in terms of the 'organization of urban space and architectural technique', however, did not take place until some time toward the end of the 7th century BC. It is then that there is evidence of the first forum, of permanent homes built of stone with tiled roofs from 625 BC, of the first building on the site of the Regia, an archaic temple on the Capitol, and other public buildings and monumental architecture [Cornell, 1995, p48, p57, pp93-94, p103, p126]. Only at this point one could say that the city was properly founded – but by whom?

It seems likely that these developments could all be attributed to the Etruscans. When the Italic-speakers chanced upon the Tiber, it is more probable that Rome was already part of the Etruscan empire. From the 8th to the 5th centuries BC the Etruscan empire was at its height, with a territory referred to as Etruria that included many city states [Cornell, 1995, p45], stretching from the Po valley in the north – where they came into conflict with the Celts – to Campania south of Rome [Berresford Ellis, 1990, p25].

The Etruscans, who had close links with the Phoenicians,[24] were drawn to the Tiber for reasons of trade as much as anything. The site of Rome lay on convenient trade routes. It had a natural ford for crossing the Tiber, easy connections via the river to the sea, and a well-established road, the Via Salaria, which connected the interior with salt beds at the mouth of the river. Some of the earliest sanctuaries have joint inscriptions in Etruscan and Phoenician, thus confirming the presence of the Phoenicians and the importance of trade [Cornell, 1995, p48, p112]. Furthermore, Phoenician amphorae are to be found in the graves of the wealthy élite.

The Etruscans also had close trading relations with the Greeks, especially the Euboeans. Their closeness is evident in Greek loan words in Etruscan of 'technical terms for vases and drinking vessels', indicating a Greek influence in 'feasting and guest-friendship culture'. The Greeks brought an oriental influence into the western side of the Italian peninsula, as they imported luxury goods from Egypt, Palestine, Syria and Mesopotamia through the Levant. From the period 730-580 BC hundreds of tombs in Etruria, Latium and Campania contain evidence of this trade from the east [Cornell, 1995, p89, p81, pp85-6]. The Greeks brought more than trade to the Rome area, influencing both religion and politics too.

Toward the end of the 7th century BC it is possible to see Rome beginning to take on some of the features of the Greek polis and, from the 6th century BC, the monarchy in Rome even resembled the Greek tyrannical model [Cornell, 1995, p118, p145]. The Tarquin dynasty – believed by Cicero to have descended originally from Corinthian Greeks[25] – ruled Rome with extreme cruelty in the 6th century BC and, like their Greek counterparts, alienated the aristocracy by distributing Rome's wealth among the populace, a policy which no doubt stimulated feelings of revenge and contributed to the end of the monarchy and the creation of republics in 509 BC [Cornell, 1995, p148].

Latin Infiltration

It is hard to date the arrival of the Romans in Rome because at no point is there any evidence of a sudden invasion. Literary sources identify three Latin tribes who supplied cavalry and men for fighting as early as the 7th century BC. It is also around 700 BC that the earliest Latin inscriptions began to appear in an alphabet script used to represent Etruscan, but based on a Greek adaptation of Phoenician signs [Cornell, 1995, p114, p103].

Since the Latins do not appear to have attacked the Etruscans and then set up their own city, the process seems to have been a more gradual one of infiltration, adoption of the host nation's customs and eventual domination of it – like the proverbial cuckoo in the nest – possibly starting in the 7th century BC. The reason the Romans were able to take over the Etruscans could be attributed to the Etruscans' character. They were described by their Latin neighbors as having 'a feminine and dreamy personality' [Jones, 2000, p212]. The sophistication we associate with the Romans was much more likely to have been acquired from the Etruscans.

Thus, from the early years of the monarchy in the 7th century BC onward – long after Rome became Roman and the creation of the republics in 500 BC – the holders of important Roman offices, whether kings or magistrates, continued to have the external trappings of Etruscan royalty in their dress and insignia. The purple-embroidered robes, the *toga purpurea* (no doubt acquired from the Phoenicians, famous for purple-dyed fabric), the *tunica palmate*, the chariot, the gold crown and ivory scepter surmounted by an eagle were all Etruscan, as were the purple bordered robe, the toga *praetexta* of the magistrates, the *sella cuvulis* or folding chair of ivory, and the *fasces*, the bundles of rods and axes carried by lectors to symbolize magistrates' power to inflict punishment [Cornell, 1995, p165].

Even after Rome became definitively Roman, some time in the 6th century BC, the Romans continued to use Etruscan rituals in the foundation of colonies or cities [Cornell, 1995, p166]. Likewise, the predominant architectural style of sacred and public buildings remained Etruscan, or Tuscan, for a long time. This included the first Roman temple, the great Capitoline Temple, built by the Etruscans toward the end of the 6th century BC because they had the skills. This first

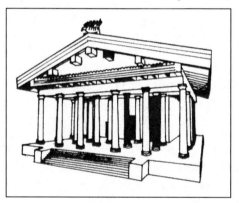

Reconstruction of the Capitoline temple in Rome

distinctly Roman temple was dedicated to the cults of Jupiter, Juno and Minerva, who were not Etruscan deities.

Rome becomes Roman

It is with greater certainty that one can point to the presence in Italy of the Latins in the 6th century BC. From this time they began to assert themselves within Etruscan society. For one thing, around 580 BC there is a noticeable change in burial customs. Luxury items disappear from Latin tombs, replaced by simpler graves without grave goods. At the same time there was a greater emphasis on public display at funeral rituals. Only the Etruscans continued with elaborate tombs in the 6th and 5th centuries BC [Cornell, 1995, p106, p172].

Other building work that occurred in the second half of the 6th century BC could well have been at the instigation of the Latins, even if designed and executed by the Etruscans. Around 530-520 BC houses were built on the northern slope of the Palatine around a central courtyard with a basin underneath to catch the rain – such water management being a distinct feature of the civilization model. Thatched huts were gradually replaced with stone buildings and there was an increasing appearance of sanctuaries dedicated to Roman gods and goddesses such as Diana or Castor.

By 500 BC Rome had become a great city, a powerful 'showplace of the western Mediterranean' [Cornell, 1995, p96, p209]. By 500 BC Tarquinius Superbus, the last of the Roman kings, had been overthrown and Rome had become a republic with two annually elected consuls. Rome was a major force in Italy to the extent that the city could count on the use of an army of 6,000 hoplites [Cornell, 1995, p189]. It was another hundred years before the Romans began to flex their muscles, appropriating Etruscan territory by invading Etruria and capturing Veii in 396 BC, nearly doubling in size the land controlled by Rome. This first expansion of Roman territory marked the beginning of a military empire that lasted nearly 1,000 years.

During this time Etruscans were still part of Roman life: names that are clearly Etruscan can be found on the lists of consuls long after Rome became a republic. Nevertheless after 500 BC the Etruscans became increasingly marginalized in the first three centuries of the republican era. The Romans began to see them as backward and old-fashioned. Etruscan terms that remained in Latin were regarded as technical and largely archaic [Cornell, 1995, p169]. Whereas the Romans went forward and became a dynamic and powerful force in the ancient world, the Etruscans ossified and eventually disappeared, their language continuing to be spoken only until the 1st century BC.

From the 5th century BC the history of Rome then becomes one of constant warfare, *bellum perpetuam*, with a high proportion of the citizenry in military service. Historian T J Cornell describes the rhythmic pattern of annual warmongering, with spring and autumn rituals to mark the beginning and end of campaigning, as being characteristic of Italic society in the 5th century BC. He points out that, by the end of the 4th century BC, when Rome had become identifiably a slave society, 'war and conquest both created and satisfied the demand for slaves', not least because the use of slaves also freed more males for military service [Cornell, 1995, p308, p333].

Admittedly, there was a temporary setback when, six years after Veii, the Celts managed to sack Rome in 390 BC, defeating a 40,000 strong Roman army. They withdrew only on payment of 1,000lbs in gold; and 'for the next 50 years, the Celts continued to harass Rome' [Berresford Ellis, 1990, p27, p30]. In response to this attack the Romans then constructed a new city wall of ashlar masonry, extending over 11km and enclosing an area of 426 hectares [Cornell, 1995, p320]. After that the city of Rome became a defensible base, thus making the most of having seven hills surrounding a natural valley. The Romans could attack and dominate the surrounds, and eventually the rest of the known world, from those hills.

The Power of Rome

The Romans were undoubtedly warlike. The name 'Rome' even comes from a Greek word for strength (Ροµι) [Cornell, 1995, p320]. It was not brute strength alone which explains how Rome managed to become so powerful. The Greeks, the Celts and many other tribes from the Pontic-Caspian and beyond, were equally pugnacious.

The Romans admired the Celts for their individual heroism – as demonstrated in a moving Roman tribute to their bravery, the statue of the dying Celt. This statue, now in the Museum of the Capitoline, is of a Celt who is naked, save for a torque around his neck, sitting on his shield with his head bowed as his lifeblood ebbs away. The Celts were quite happy to decide the outcome of a battle either through single combat to the death between hand-picked individual warriors, or through long lines of naked warriors with only gold torques around their necks. The Celts fought naked 'for religious reasons' – although there is no doubt thousands of men '...of splendid physique and in the prime of life', wearing only gold torques could have been rather intimidating, even for Romans! [Polybius, quoted in Berresford Ellis, 1990, p38]

No doubt part of the reason for the Roman success was their particular fighting style and, unlike the Celts, the willingness of individual soldiers to be subsumed within a fighting unit. In contrast to the Celts and the Greeks, with the possible exception of Sparta, the Romans had a different, militaristic way of thinking apparent in their approach to colonial expansion. When, in 334 BC, they captured Cales on the main route from Rome to Capua, 2,500 families from the Roman proletariat were taken to live there, as well as a smaller, elite group which was given land [Berresford Ellis, 1990, p352]. Before that, in 381 BC, the Romans had allowed another captured territory, Tusculum, to retain some autonomy, becoming the first Roman *municipium*, provided that Tusculum fulfilled its obligations to service in the Roman army. But when Tusculum joined a revolt against Rome in 340 BC,[26] Cales then became the preferred model for colonial expansion.

The Romans had a clear idea that civilization concerned the establishment of cities. The building of cities and living as citizens was not fundamentally of interest to Celts, even though Celtic settlements were the foundations of many great cities throughout Western Europe. The imposition of a form of civilization on a local population was not the Celtic way: they were more likely to build hill-forts, creating farming settlements and cohabiting with the local population [Berresford Ellis, 1990, p94]. The Cisalpine Gauls, living south of the Alps, were typical in being more interested in farming than warfare. They were an advanced agricultural and pastoral people when they settled in the Po Valley. The archaeological record has plenty of evidence of Celtic agriculture – particularly growing wheat and barley, keeping livestock and other domesticated animals. It is thought that the wool trade from the Celts, including the making of carpets, generally 'clothed the greater part of Italy' [Berresford Ellis, 1998, p31].

The Romans shared their attitude to cities, and to war, with the Greeks. Asia Minor is littered with the remains of the Greek *polis*. For both societies war was a way of life. War was the means by which they acquired booty in the form of transferable goods, slaves, livestock and land. The Romans shared the Greeks' Homeric noble ideal which regarded trade as being associated with low-status foreigners, 'Sidonians' (Phoenicians), who were thought untrustworthy and profit-seeking. The Romans believed that the best way to acquire goods was through 'looting and piracy' [Aubet, 1993, p103].

Roman Success

In spite of these similarities, the Romans were still ultimately more successful than the Greeks. The Romans succeeded where the others failed, above all, because of their unique ability, sustained over an exceptionally long period of time, to combine discipline and organization with opportunism and flexibility. To use a rather hackneyed analogy, if ancient Rome had been a modern business corporation with a stock market quotation it would definitely have been a 'buy'. In another way of describing them, it could be said that they combined some of the more acceptable attributes of the German Nazi Party with some of the many positive aspects of the USA – though these comparisons should not be taken too far. But is it so surprising that all three share the emblem of the eagle? [Berresford Ellis, 1990, p25].

In common with the Nazis, the Romans had an ideology based on the concept of personal sacrifice, not to the party as in the case of the Nazis, but to the city itself, for the glory of Rome. This uncomplicated objective was a powerful unifying force, especially after the demise of the monarchy and the rise of the republics, because it encouraged a ruthless and unquestioning belief that the end – the glory of Rome – justified the means. It was acceptable to put to death every tenth man (the literal meaning of to 'decimate') in an army unit in order to maintain discipline. The Romans were willing to sacrifice free will and submit themselves to a level of control with which we would have difficulty today.

Without discipline and organization, the desire to unite around a common goal would, however, have had little lasting effect. One aspect of the civilization model that the Romans took extremely seriously was the importance of a legal framework for society. In the middle of the 5th century BC when Rome was experiencing trouble with the plebeians refusing to perform military service, the crisis was resolved through power-sharing agreements and yielding to plebeian pressure for the codification of laws to protect them against arbitrariness. It was also around this time that the Romans sent men to study law in Athens, also appointing the *decemviri*, ten men, to govern and draft laws [Cornell, 1995, p257, p265, p272]. Roman meticulousness in legal drafting continues to form the basis for many legal codes today.

Another characteristic that they shared with the Nazis was their attitude to social stratification and a bureaucratic attention to detail. This Roman obsession with record-keeping was made easier by the decision, probably taken very early on, to base their social and military organization on units or multiples of ten. Whether or not they gained this idea from the Phoenicians, who had a Council of Ten, it is hard to say [Aubet, 1993, p120].

From the middle of the 7th century BC, it is said that three tribes, who supplied cavalry and infantry, formed the *curiae* with ten members each. These *curiae* had important functions in daily and religious life and were the foundation of urban administration, meeting as the *curia comitata*. Each *curia* then provided 100 men, the centuries, as the basic formation of the army [Cornell, 1995, p114, p117]. King Servius Tullius was able to take advantage of the structure provided by the *curia comitata* when he decided to reform the army in the 6th century BC, by first undertaking the radical step of conducting a census.

Servius Tullius wanted to change the army so that the Romans could emulate the Greek style of fighting, using hoplites. This style had appeared in the Greek world around 700 BC, becoming a standard Greek method of warfare from 675 BC onward, and spreading to the western Italian peninsula around 625 BC. Using the round shield of the hoplite to create massed phalanxes required greater discipline than other forms of warfare, as it was important for individual soldiers to hold their place in the line [Cornell, 1995, p184].

Social Stratification

Servius Tullius wanted to know who in Roman society was available and rich enough to fight with this new Greek method using heavy armor. To that end he conducted the first census, dividing people by wealth and property into rank and status. Ironically, he was able to instigate the census as a five-yearly ritual thanks to the existence of a semi-literate society because the despised traders, the Phoenicians, had given the Romans their alphabet and written script, thus also giving them the means to further develop their society.

The purpose of Servius' census was thus to classify all men fit for military service. Indeed, the word for 'class' is based on the root word *calare*, meaning to 'call or summon'. He then divided each class by age into smaller groups so that each one contained equal numbers of centuries of *iuniores* (men aged between 17 and 45) and *seniores* (aged 45-60). The 'juniors' were expected to become front-line soldiers and the 'seniors' the home guard. The groups were then differentiated by armor depending on their place in hierarchy.

Servius also needed to establish who was financially able to carry out military service. In the 6th century BC there was no concept of a state-funded warrior class and only those who could afford it could join the army. In 500 BC Romans still had warlords with private armies, some with as many as 5,000 men. Pay only began to be introduced nearly 200 years later as military operations became more professional.

At the end of the 5th century BC Livy records wages being paid to soldiers (the *stipendium*) to compensate them for loss of income during prolonged campaigns. By the beginning of the 4th century BC a property tax (the *tributum*) was imposed in order to pay for military expenditure, which included the increasing tendency to requisition supplies for the army [Cornell, 1995, p173, p179, pp183-4, pp143-4, p186, p188, p313].

The classification of Roman society in the manner of the Servian reforms had several benefits. The census both reflected society's stratification and enforced it. It inadvertently gave Roman society the means to transform itself and, in one sense, these reforms were the monarchy's undoing. Because the centuries were in effect a cross-section of society, based on age and not tribe, they had the advantage of diminishing the aristocracy's power base within the army while, at the same time, increasing the power of the central state. The centuries were also the means by which levies (the *legio* or legion) could be raised to recruit for the army [Cornell, 1995, p182, p194].

Servius further consolidated the power of the centuries by replacing the *curia comitata* with the *centuriate comitata*. The *centuriate comitata* became, to all intents and purposes, a people's court with a right of appeal, fulfilling legal and political functions as well as military, but one also in which the voting was weighted so that the old could outvote the young and the rich could outvote the poor. When Rome's patrician families had had enough of the tyranny of the later Tarquinian dynasty, the *comitia centuriata* was the obvious structure to replace the monarchy and facilitate the establishment of a republic, as it was the *comitia centuriata* that then elected the consuls that ruled Rome [Cornell, 1995, p196, p185, p226].

Roman Republicanism

It could be said that the institution of a republic was a Roman innovation since it was not based on either the Greek model of democracy nor was it an oligarchy restricted to a few wealthy individuals. Another advantage of the regular use of the census was that it became a mechanism whereby the powerbase of the Republic could renew itself. Thus, 'at each census the city-state was reconstituted' every five years in the classical Republic, in a ceremony around the Campus Martius, effectively re-founding the city. Here the symbolic membership of the centuries was reconsidered each time, using a system of lots [Cornell, 1995, p191].

This ability to be flexible is no doubt one reason why the Romans actually did successfully create a 1,000 year 'reich', unlike the Nazis, whose empire was brief. The Romans combined flexibility with opportunism and an openness to change – which is what the Romans had in common with the United States of America. They also did not suffer from the constraints of the Nazis' absurd notions of racial purity.

Just like modern America, anyone could be accepted as Roman if they qualified. For instance, famous Romans like Virgil were actually Celts. Rome's foundation myth, which described Romulus actively seeking colonists, reinforced the idea that anyone could join the Roman adventure. It was an ideological message that promoted the belief that Romans were a mix of different ethnic groups united by a shared subscription to the principles of *romanitas*. By the beginning of the 3rd century BC the appeal of the *Aeneid* as an epic poem had spread; the story of the twins and the wolf had become standard and began to appear on Roman coinage. In addition, Rome was 'unique among ancient

societies in its practice of assimilating freed slaves, who automatically became Roman citizens on manumission'. By the end of the Republic many aristocrats had servile blood and a large proportion of the city population were either slaves or freedmen [Cornell, 1995, pp60-61].

statue of Romulus & Remus with the she-wolf

Roman flexibility was severely tested, however, in the middle of the 5th century BC when faced with revolt by the plebeians. In 494 BC a group of them, 'oppressed by debt', withdrew to an area outside the city where they formed their own organization and refused to do military service. In an attempt to placate them in 447 BC two *quaestors* were elected to assist the consuls. But the plebeians were not reintegrated into Roman society until 367 BC when the *nobilitas* was formed, a new creation of a joint plebeian-patrician, power-sharing aristocracy [Cornell, 1995, p256].

This ability to create social mobility allowed for the influx of new ideas. It meant that the Romans did not suffer from the later English disease of 'not invented here'. They particularly liked taking a good idea from the Greeks and beating them at their own game. It mattered little to the Romans whether an idea came from the Greeks, the Etruscans, the Phoenicians or the Celts.

A lot of what we take for granted as being Roman, even in matters of warfare, often came from the Celts. The Romans and the Etruscans adopted the Celtic style of helmet, a round cap with a knob on top. The Romans took on the Celtic long shield and it is possible that the Latin word *scutum* is a loan word from Celtic. The Celtic word for spear, *lancea*, appears in Latin. The battle tactic of *testudo*, or tortoise, of interlocking shields, was a Celtic practice [Berresford Ellis, 1998, p88, p91, p102].

The Celts were also renowned horsemen and Celtic cavalry became incorporated into the Roman army. Words like *equus*, Latin for horse, were originally Celtic; *caballus*, a Latin word found at the root of words like 'cavalry', was Celtic for 'pack horse'. Further Celtic influence on Roman transport can be found in a rich vocabulary of Celtic loan-words such *carpentum*, Latin for a two-wheeled baggage wagon. It has at its root the Celtic word *carbanto*. Most surprising is the word 'league', which we associate so closely with the Romans, deriving from the Celtic *leuga* and serving as an indication that the highly mobile Celts built extensive road systems in Europe long before the arrival of the Romans [Berresford Ellis, 1998, pp94-99].

End of Egyptian Paganism

While this brief survey of certain characteristics might give some idea as to how Rome came to be such a dominant force in the ancient world, it does not explain how or why the Romans came to deliver the final blow to Egypt in the 4th century AD. After all, Egypt had been under Roman control for more than 300 years before the massacre on Philae Island. So, what changed?

The simple answer is that the Romans converted to Christianity. Therefore the conventional answer as to why the Romans killed the priests at Philae, having closed all the other temples in Egypt, was that they were now doing the 'right thing' in the name of the Christian God by ridding the world of pagans. Even the use of the word *pagan* ('villagers') as a term of abuse was indicative of this change in attitude. Those who continued believing in the old religion were essentially ignorant country peasants, not city-based civilizers. Civilization had found the 'true' religion in Christianity and was finally truly 'civilized'. We now had grain storage, sewage systems and the modern idea of One God. But how did such a powerful society as Rome become so vulnerable to this new religion? Was it really a case of 'seeing the light'?

Chapter 15

Rome Embraces Christianity

Why should monotheism have been of interest to a military society like the Romans? After all, they were more interested in capturing booty than in saving souls. How could the Romans, with the most powerful empire in the world, have allowed Catholicism to become so powerful that even the Emperor could be excommunicated? What was the appeal of a religion that in the end rejected their entire belief system, dismissing it as pagan nonsense?

The conventional view, that as pagans they realized the error of their ways and so converted to the 'true' religion, does not really stack up. The Romans who contributed to the destruction of Rome under Christianity must have been very different from those who built Rome. But was there also something deep in the Roman psyche that enabled Catholicism to take a hold?

Certainly, from at least the 2nd century AD onward, there were well-placed, intelligent Romans and Greeks, such as Suetonius, Epictetus, Marcus Aurelius, Celsus and Porphyry, who had nothing but disdain for the new religion [Hoffmann, 1987, pp27-28]. Suetonius thought Christians to be superstitious. Epictetus believed that they were driven to martyrdom by 'blind fanaticism'. Marcus Aurelius, the emperor with a deep interest in Greek Stoicism, had enormous antipathy toward Christians and regarded their enthusiasm for martyrdom as 'exhibitionism' [Frend, 1984, p170]. There was no inevitability about martyrdom: Roman officials 'often tried to persuade the accused [the Christians] to save their own lives' just by asking them to 'swear by the emperor's genius', which they would refuse to do and so were condemned [Pagels, 1979, pp96-97].

Celsus, the 2nd century AD Greek commentator, was puzzled by the many inconsistencies of the Christian faith. What he and others struggled with, for example, was the Year Zero, the *tabula rasa*, aspect of Christianity and its appeal for sinners. Celsus could not understand why God would suddenly sacrifice his only son; he wondered why God had waited so long to judge men, and asked 'Did he not care before?' [Hoffmann, 1987, p77]. In his opinion, God did 'not need to inflict correction on the world as if he were an unskilled laborer who is incapable of building something properly first time around', and he thought it arrogant of Christians to believe that there was more evil in the world before Jesus [Hoffmann, 1987, p82].

We can at least read Celsus' comments by reconstructing his *Discourse against the Christians* from the polemical reply written by the 3rd century theologian Origen – one Christian for whom 'veneration of martyrdom [was] the ultimate expression of Christian commitment'. Porphyry, however, wrote 15 books against the Christians in 3rd century, all of which were burnt by the emperors Theodosius II and Valentinian III in 448 AD [Hoffmann, 1987, p29].

The Force of Fanaticism

How did the relatively small, self-important group of Christian religious fanatics manage so successfully to infiltrate such a powerful empire as the Roman Empire; thereby affecting the course of Western intellectual development over the next 1,600 years? There are parallels with more recent history: we have only to witness the impact of ideologies such as Communism or militant Islam to see the force of single-minded belief and its impact on populations. Early Christians similarly benefited from the cohesive effect of a rigid belief-system with a concept of sacrifice at its core – the canon of belief having been fairly well established by the Council of Nicaea.

Ironically, it was a pagan Roman emperor, Constantine, who

did much to unite the early Church into a powerful force. From a political point of view, Constantine could see that Christianity could have a unifying effect on the empire – all could join together to fight under the banner of One God, singing from the same songsheet. Constantine himself remained ambivalent about the new religion and did not actually convert until he was on his deathbed. He continued to have *Sol Invicto Comiti* on his coins [Frend, 1984, p484], and in 321 AD decreed that Sunday should be kept as a day of rest for the 'veneration of the Sun' [Frend, 1984, p488; Brox, 1994 p105]. At the same time, he actively encouraged the new religion.

The Church had been split with dissent. From the start there had been the differences between the Gnostics and the 'literalists'. Then there was a row which had started in the 2nd century AD over the peculiar question of whether Jesus was *of* God or *from* God. As Son of God, was he a man or a part of God? Constantine decided to intervene, calling the Council of Nicaea in 325 AD, where he forced a compromise that stated that Jesus was 'consubstantial' (*homoousious*) with God. This became the basis of the Nicene Creed, or statement of belief [Frend, 1984, p499].

Another factor was who they were, who they recruited to join the Church and how they did it. Without doubt, fanaticism – religious or otherwise – in itself has an enormous psychological appeal to a certain sort of person and early Christians were clearly different from the rest of Roman society. They kept themselves apart. They were 'accused of enjoying the benefits of society without sharing its political burdens', even though they claimed loyalty to the emperor. They were criticized for not taking their responsibilities in society seriously, as they refused to take part in military service. They did not make sense to other Romans because they 'declared worthless all the values prized by society – science, education, culture, possessions, career...'. They were the original killjoys, avoiding popular festivals, theatre plays and games in the circus. An even worse accusation was that they split families [Brox, 1994, p33-35].

As far as possible, the early Church 'regulated access to Christianity with great care and strictness'. The Church excluded anybody who might question its tenets, such as 'actors, sculptors and teachers, because they lived on pagan belief in gods or taught it' [Brox, 1994, p95]. Celsus and others thought that Christians did this because they 'do not want to give or receive a reason for what they believe', and that converts were 'not to ask questions but to have faith'. He accepted that 'a few moderate, reasonable, and intelligent people... are inclined to interpret its beliefs allegorically' [Hoffmann, 1987, pp27-28, p57].

Otherwise the Christians' attitude was, according to Celsus, 'Let no one educated, no one wise, no one sensible draw near. For these abilities are thought by us to be evils. But as for anyone ignorant, anyone stupid, anyone uneducated, anyone childish, let him come boldly'. He accused them of telling children 'not to believe in their fathers or their teachers', and of telling their own pupils that 'too much learning is a dangerous thing: knowledge is a disease for the soul, and the soul that acquires knowledge will perish' [Hoffmann, 1987, p75].

Celsus claimed the Christians deliberately wanted 'and are able to convince only the foolish, dishonorable and stupid, and only slaves, women and little children'. He described how, in the marketplace, Christians would avoid discussions with 'wise men' and instead were surrounded by slaves and 'the most illiterate country bumpkins'. Lucian, a Syrian from Samosata, writing in the 2nd century AD, also 'regarded Christianity as a form of sophistry aimed at an unusually gullible class of people'. Celsus claimed that Christians would 'flee in all directions' when they saw schoolteachers or fathers coming with the excuse that 'they do not want to have anything to do with men as corrupt as these pagans' [Hoffmann, 1987, pp72-73, p25, pp72-73].

Christianity crept in, as it were, through the 'back door', and it is true that women and domestic servants were often the first to convert. In the 3rd century AD 'ladies of wealth' socially cultivated their relationships with bishops [Chadwick, 1993, p164]. The roles of Constantine's wife and mother in the 4th century were examples of this female interest. His wife Fausta donated her palace on the Lateran to the Church in 313 AD [Frend, 1984, p487], and his mother started a trend for pilgrimage lasting a 1,000 years, with visits to the Holy Land to find pieces of the 'True Cross' and other relics. Fausta and her mother-in-law both encouraged Constantine in his pro-Christian attitude.

In spite of the influence of certain high-ranking women, Christianity took root in a society in which women were almost entirely disempowered. There were only brief periods when Roman women experienced emancipation and self-determination [Pagels, 1979, pp83-84]. As a military fighting machine Rome was not exactly in touch with the intuitive, feminine side of its personality. The early history of Rome implies that this imbalance was present from the start, with the story of the rape of the Sabine women suggesting that the original Romans arrived as a male warrior sodality without females.

Social attitudes to women were reflected in attitudes to religion: Roman goddesses did not command the same respect and place in Roman life as, for instance, Egyptian goddesses like Isis or Hathor. Even Venus, the Roman goddess of love, was as often as not associated with Mars, Roman god of war. Women were excluded from participation in Mithraism, one of the most popular Roman cults, which must have been one reason for the appeal of Christianity to them.

Romans and Religion

Apart from Constantine and his family's high profile interest, an important factor in the successful infiltration of Christianity was the attitude of the Romans themselves toward religion. By the time they had a more favorable attitude toward Christianity in the 4th century AD, their own religion had absorbed so many different cults and creeds that it was no longer recognizable in its earlier form. Christianity was, in some senses, a cult that the emperor decided was more useful than the others. Romans were able to accept Christianity because, as with other cults, they thought they could mould it to their needs.

They were no more concerned with the 'truth' of Christianity than of any other religion they came across. Their approach to religion was essentially pragmatic and rooted in natural phenomena. From the beginning, from the Pontic-Caspian, they had inherited a hierarchy of deities with a sky-father god at the head of it – Jupiter, who threw thunderbolts.

From very early on in the development of Rome, a Greek religious presence is apparent. One of the earliest items

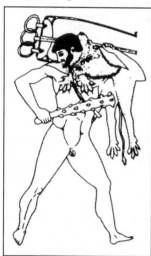

Hercules

discovered at the shrine of the *Niger Lapis* in the Roman *Comitium* is a fragment of a Greek black figure cup dating 580-570 BC, showing the Greek god Hephaestus, who equates with the Roman god Vulcan, returning to Mount Olympus on a donkey. As in other matters, the Romans were happy to use Greek examples to create their own versions of deities. They adopted the Greek god Heracles as Hercules, whose appeal to the Romans was predictable, given his reputation for trials of strength. Life-size statues of Hercules and Minerva dating from 530 BC have been found at the Temple of Sant'Omobono, itself near the site of an ancient cult of Hercules and closely associated with river trade [Cornell, 1995, p162].

The Romans do not seem to have had their own temples until after that date, with the great Temple of Jupiter on the Capitoline being built at the end of the 6th century BC. Trading centers on the coast and along the River Tiber were another means by which different cults arrived in Rome, with traders often building their own sanctuaries. Given the strong trading links between Rome and the East through the Phoenicians, there was for a long time a strong oriental influence on Rome. As early as 460 BC, there was an Etrusco-Italic temple at Pyrgi, one of the harbors of the Caere, containing a dedication to the Phoenician goddess Astarte.

They continued to accept the religion of others – dedicating, for instance, a temple to the Greek god of healing, Aesclepius, on the island in the Tiber in 291 BC [Cornell, 1995, p397]. Later on, during the 1st century BC, Phoenician ideas of the sacred marriage between the king and a goddess were in evidence in Rome, and the tradition arose of the Roman king Servius Tullius having a love affair with the goddess, entering her window at night [Cornell, 1995, p112, p146].

As the empire expanded through military conquest, so the Romans came into contact with other religious ideas. The Celts were the sworn enemies of the Romans, yet the Romans had no difficulty in accepting the Celtic horse goddess Epona into official Roman religion between the 1st and the 4th centuries AD [Berresford Ellis, 1998, pp99-100]. The religion with the most impact on Rome from about 170-240 AD, however, was the cult of Mithras [Frend, 1984, pp276-8].

Mithras slaying the bull

Of Persian origin, Mithraism was especially popular among the Roman military because it celebrated the prowess of Mithras as a heroic figure, similar to Hercules. After slaying the sacrificial bull, Mithras carries it on his shoulders [Gilbert, 1996, p55]. Not surprisingly, elements of

Mithraism continued to shape the Roman interpretation of Christianity long into the 4th century AD.

The special buildings, the Mithraea, in which Mithraic rituals were enacted, became the model for the first Christian churches. These were small narrow buildings with a central aisle and seats for 20-25 on either side. At the end was an altar block with a carving of Mithras slaying the bull on one side. Central to the ritual was the sacred meal, with similarities to the Eucharist, that Mithras consumed with the Sun God, which the participants celebrated with bread and wine [Frend, 1984, p277].

Initiation

Regardless of its popularity, Mithraism, in common with the Greek mystery religions, still had one major drawback: initiatory barriers to entry. Part of the powerful attraction of Christianity must have been that it was easily accessible, with only baptism and the Eucharist as the main sacraments [Brox, 1994, p93], and no complicated initiation process or mystery rite that had to be kept secret.[27] Whoever wanted to join spent time as a *catechumen* before baptism, but once baptized the new member was then accepted as part of the community.

Unlike paganism, the early Church had no sacrifices, no temples and no altars. Instead, the early Christians used houses as churches, with tables for the Eucharist which had 'neither the form nor the function of altars'. Not only did the new religion embrace almost anybody (with the exception of professional teachers, actors or anyone likely to question its tenets), but the more powerful and sinful, the better.

By contrast, worshippers of Mithras were concerned with acts that helped the believer to 'move through the seven spheres which surrounded the Earth, to salvation'. The Mithraneum in Ostia, for instance, contained seven doors symbolizing this process. The initiate was blindfolded by the master of novices, sworn to secrecy and reborn at the lowest of seven grades of perfection [Frend, 1984, p277]. Even more problematic for the Romans were the mystery religions.

The mystery religions, as practiced most famously at the school at Eleusis on the coast near Athens, would accept for initiation only those 'pure in heart'; those who had, for example, not committed murder. This was an impossible stipulation for certain emperors, such as Nero, who were denied participation in the Eleusian rites.

This ethical difference between Christianity and the mystery religions nonplussed Classical commentators. While the moral philosophy of the Greeks emphasized the importance of being 'good', none of the Classical commentators could see why Christians would want 'whoever is a sinner' and not the 'pure in heart' [Hoffmann, 1987, p74]. It made no sense to them that Christians were interested only in the sinful. Surely everyone was at least a little bit sinful, so why was there no general call for salvation? Celsus was not alone in thinking that Christianity appeared prejudiced against 'virtuous' people.

While the ethical aspect of the mystery religions provided one difficulty for the Romans, the transcendental nature of the mystery schools' teachings was another. The Romans had no understanding of and no interest in a shamanic-style engagement with the spirit world. They had never had their own tradition of alchemy. They remained a militaristic society on the periphery of true civilization (i.e. Egypt). They were never taught a Roman equivalent of the Egyptian *Heb Sed* festival and the Osirian rites.

In the 300 years that the Greeks ruled Egypt they made themselves acceptable to the Egyptians. In 330 BC Alexander the Great had gone to Egypt with the express intention of learning more about Egyptian religion. The Greek pharaohs, the Ptolemies [Osman, 1998, p277], had then developed a hybrid Greco-Egyptian religion that absorbed Osiris into Serapis. On mainland Greece, the Greeks formed their own version of the

Egyptian Osirian rites with the symbolic death of the Greek god Dionysius. Even when the Romans successfully invaded Egypt, after the death of Cleopatra in 30 BC, they still failed to access its inner secrets.

The Royal and Religious

The Romans were unable to carry on the traditions of the Egyptians because, unlike the Greeks in Egypt, they misunderstood the substance of the relationship between the royal and the religious, in spite of inheriting certain basic concepts of civilization thousands of years before on the Pontic-Caspian. They lacked a priesthood that could administer the trance-inducing substances and a king who was prepared to sacrifice himself by undertaking risky out-of-body experiences on behalf of his people.

Without the restraint of being in the service to the people, it was easier for kings to become tyrants. The role of the priesthood was reduced to administering pointless ritual sacrifices and setting the calendar – one of the ancient functions of the priesthood that the Romans implemented from the beginning, from at least the 6th century BC.

There was little attempt to interact with the Divine on the part of the Roman priesthood, in the sense of the Hermetic dictum of 'as above, so below'. Even the practice of basing divination on the innards of sacrificial animals 'remained the preserve of special Etruscan priests called *haruspices*' and was 'always a distinct and marginal area of Roman religious life'. It was not recorded as a regular event until the Second Punic War when the *haruspices* were summoned from Etruria especially to interpret signs [Cornell, 1995, pp165-66].

Nevertheless, the Romans did retain an echo of the original model of civilization in one particular priestly role. They paid lip-service to the concept of a 'priest-king', the *rex sacrorum*, described as 'a priest whose task it was to perform the religious functions of the erstwhile king'. So insignificant was the status of the *rex sacrorum* that he was even chosen by another priest, the *pontifex maximus* – whose title connects with the later institution of the Pope, who continues to be known as the pontiff [Cornell, 1995, p233].

The *pontifex maximus* and the *rex sacrorum* performed sacred tasks together in the rebuilt *Regia*, which contained the shrines of Mars and Ops Consiva, the provider of wealth – the *Regia* having been burnt down in the difficult transition from monarchy to Republic in about 500 BC [Cornell, 1995, p237]. Although the institution of the *rex sacrorum* continued to exist in the later Republic, it became increasingly marginalized and obscure. The *rex sacrorum* had no real power because, according to ancient historian Livy, he was prohibited from holding political office and membership of the Senate. His role was limited to the operation of the calendar, the performing sacrifices on the first day of each month (the *Kalends*), and announcing the dates of the months' festivals on the *Nones* (nine days before each full moon) [Cornell, 1995, p105].

The inability of Roman priests to access metaphysical knowledge central to civilization is evident in the physical remains of their culture. One cannot fail to be impressed with the high standard of Roman masonry and engineering (the Pont du Gard in the South of France is a breathtaking example).[28] But, because they did not possess the secrets of the ancients, Roman building lacks certain key identifiers of the ancient model, such as the use of megalithic, polygonal stone in the building of walls.[29] They were as likely to use brick in the place of stone in the construction of some of their most impressive buildings, since they did not know how to cut, dress and move really enormous pieces. They had to invent the pulley because they had no other means of lifting large pieces of masonry. Their use of the arch was an elegant solution to their inability to span large spaces by any other method. They also had to use mortar in construction work; whereas the fine stonework of the civilizers had no mortar between joints.

Good Christians – Bad Pagans

The Romans must have been aware that knowledge belonging to the original model of civilization eluded them. Given some of the campaigns they fought and the groups they persecuted, they must have felt threatened by its existence beyond their grasp. Initially unable to conquer Egypt, the Romans seem to have decided to destroy pockets of ancient knowledge wherever it existed.

We are given the impression that only Christians were persecuted under the Romans – only they suffered martyrdom. That was not the case. In spite of their acceptance of some religious cults, Romans were not universally tolerant. 'Bad' Roman pagans did not suddenly see the error of their ways and stop martyring the 'good' Christians in order to persecute other pagans. There was no sudden switch. In fact, the Romans had been attacking other pagans and the old religions for centuries.

In Italy, Neoplatonists and Pythagoreans were a favorite target. Tacitus refers to *mathematici*, as Pythagoreans were known, and *magi* being thrown out of the country in AD 16 [Chadwick, 1997, p72]. The Romans also banned alchemy whenever they came across it, long before Christianity associated it with paganism. Zosimus, a Syrian who lived in Panopolis on the Nile and who created an encyclopedia of alchemy, witnessed Egyptian priests in the 3rd century AD still secretly trying to preserve alchemical tradition within the temples, in the face of the oppression by the Romans. Emperor Diocletian (AD 284-305) burnt books on alchemy 100 years *before* the Romans became Christian [Baigent, 1998, pp208-9].

The Romans deliberately tried to eliminate the Druids, the Essenes and the Chaldeans – all known to be 'carriers of secrets from the past' [Velikovsky, 1978, p176]. Why else would the

Romans have invaded the areas where none of these priesthoods were a military threat – with the possible exception of Druids who were sometimes chieftains as well – in places of little strategic value to the Romans? It wasn't because they had converted to Christianity.

Chaldeans, Essenes and Druids

Thus, in 38 BC Mark Anthony invaded the Commagene [Gilbert, 1996, p218], a mountainous country in central Turkey that was of little use to the Romans. In AD 72 the emperor Vespasian deposed the Commagene king Antioch IV [Velikovsky, 1978, p175]. The only explanation for this interest in the Commagene must have been that it was the home of the Chaldeans, the famous astrologers who, by the time of the Romans, had moved north from Sanliurfa into the remains of the collapsed Hittite empire in the Commagene.

Similarly, the main objective behind Suetonius Paulinus' decision to destroy the island of Anglesey in AD 61 was that it was a popular redoubt of the Druids [Rankin, 1987, p220].[30] If the Romans had just wanted to protect the northwestern edge of the empire, they could have used the English Channel as an easily defendable border, with no reason to move the empire beyond Gaul. Apart from the tin trade from the southwestern corner of Britain, which they could have had anyway without invasion, the British Isles were of little interest to the Romans.

Moreover, there was not necessarily any strategic advantage in building Hadrian's Wall as their most northern border. From a military point of view, the wall might have been more effective along the edge of the River Tweed to the north. One plausible explanation for choosing the location of Hadrian's Wall was that this line of latitude, 55 degrees north, was important to the Druids. This was in connection with the calibration of calendars using the shadows from the sunrises and sunsets at the solstices and equinoxes [Knight & Lomas, 2000, p199].

The Druids were a threat to the Romans because they had a considerable reputation for teaching. The name 'Druid' probably comes from a root, *wid*, meaning 'knowledge' and *dru*, meaning 'very great' [Chadwick, 1997, p12]. They were regarded as natural scientists with an understanding of physics, astronomy and mathematics, as well as medicine. Pliny in the 1st century AD 'presents the Druids as doctors and magicians, or a combination of the two, dealers in unnatural natural science; he emphasizes their medical and magical practices, and their possession of magical recipes'.

Caesar held the view that the Druids taught youths 'many things respecting the stars and their motions, respecting the greatness (or size, *magnitudo*) of the world and of the Earth, respecting the nature of things'. Hippolytus referred to them in the early 3rd century AD as 'seers, prophets and magicians... [who] are said to have foretold the future by ciphers and numbers after the manner of the Pythagoreans' [Chadwick, 1997, p31, p47].

Druid teaching methods relied on committing everything to memory rather than writing, and their schools were well attended by the Celtic aristocratic youth. Therefore, one way that the Romans tried to undermine them was by creating their own schools in Gaul which did not rely on the 'art of memory'. The school at Augustodunum (Autun) in 12 BC was deliberately built on the site of a famous Druid school. These schools had the explicit purpose of attracting youths away from the Druids and making it possible for them to become Roman citizens. At the same time, the emperor Augustus introduced measures 'prohibiting druidical practices' for Roman citizens [Chadwick, 1997, pp70-71].

The Essenes shared much in common with the Druids, particularly with regard to healing. The Essenes' Egyptian successors were even known by the Greeks as the *Therapeutae* because of their reputation as healers. They were as keen on healing souls as well as bodies, using exorcism, and their physicians had a 'special gift of insight' allowing them to predict the course of the disease and the most effective treatment [Allegro, 1981, p102, p207]. The Essenes guarded precious works of the ancients on the healing of diseases using roots, 'offering protection and the properties of stones' [Dupont-Sommer, 1961, p30]. According to their *Book of Jubilees*, God allowed Noah to know 'all the medicines of their diseases, together with their seductions, how he might heal them with herbs of the Earth', which Noah wrote down [*Jubilees* X 5-14, in Allegro, 1981, p57]. In spite of this reputation, the Romans destroyed the Essenes' site at Qumran in AD 66, along with the Temple in Jerusalem, even though they had ruled Palestine directly since the 1st century BC.

Monotheism Wins

Long before Christianity, the Romans clearly differentiated between paganism that was acceptable and paganism that wasn't. Significantly, the unacceptable paganism's practitioners knew the important knowledge that related to original concept of civilization. If anything, Christianity gave the Romans a fresh motive for the destruction of the old religion – hence the massacre on the island of Philae in the 4th century AD, 300 years *after* the Romans had conquered Egypt.

As we have seen today with suicide bombers and theocratic regimes, acting 'on behalf of God' is a powerful motivational force. It made little difference to the Romans whether they attacked in the name of God or in the name of Rome. The Romans' conversion to Christianity was arguably a backward step for civilization because Christianity was a more effective strategy in a continuous process of destroying knowledge, which the Romans had begun centuries before. Using Christianity had a great impact on Western intellectual life because, as a result of their successful exploitation of the Christian story, the later Romans were able to gain control over people's minds as well as their bodies.

By the 4th century AD, Christianity as an organized religion had begun to change in a direction that was of greater appeal to the Romans. We fail to appreciate just how far Christianity had by then moved away from its original precepts, and how much the texts we take for granted have been affected by later interpretations. If we look more closely at the origins of this religion, so fundamental to Western development, we might be surprised to find that the early history of Christianity is not quite what we have been led to believe.

Chapter 16

The Unpublicized History of Early Christianity

Christianity would not necessarily have survived without the Romans. If the Romans had not decided to make Christianity their state religion, it is possible that we in the West would not have lost our polytheistic roots as comprehensively as we did. How this monotheistic religion arose in 1st century Palestine from much older Egyptian roots owes much to one person, however. Contrary to what we have been taught, that person was not someone called Jesus of Nazareth. The decision to start a religious movement known as 'Christianity' had less to do with him, and far more to do with someone else called Saul of Tarsus and his 'Damascene moment' involving the Jewish sect known as the Essenes.

The history of early Christianity hides a controversial truth. Jesus did not come from Nazareth, nor was he born at the start of first millennium AD: he may have lived a long time before or he might not have existed at all. Ancient scrolls found by chance at Qumran on the east coast of the Dead Sea in 1947 are the main reason for having doubts about the existence of Jesus. These scrolls belonged to the Essenes, the Jewish sect who occupied the site at Qumran, not far from Jerusalem. Bronze coins have been recovered from Qumran spanning a period from 135 BC to AD 136, with many concentrated in the time period AD 6-67 [Baigent & Leigh, 2001, pp233-34], thus confirming the occupation of the site throughout Jesus' presumed lifetime. The Essenes then buried their scrolls some time after AD 136. Given that the Essenes were at the site during the time that Jesus is supposed to have lived, it is reasonable to think that they might have recorded something about such a significant person.

Even though French Jesuit scholars based in Jerusalem have been translating the scrolls for over 50 years, they have been unable to establish any reference to Jesus. Not only do the Dead Sea Scrolls appear to make no mention of Jesus, there is also a view that there is 'no contemporary record, Roman or Jewish, [which] testifies to the presence of Jesus in Palestine at the beginning of the 1st century AD' [Osman, 1998, p148].

The Romans were usually thorough in their record keeping and so it is odd that they did not refer to someone of Jesus' importance as being tried before Pontius Pilot and crucified. References to Jesus in chronicles of the Jewish historian of Josephus – also possibly a contemporary – are considered to be a later interpolation to give credibility to the Jesus story. Jesus could not even have come from Nazareth, as Nazareth as a place did not even exist at that time [Baigent & Leigh, 2001, p256].

Other Jewish records have only vague references, for example in the Talmud. They refer to 'Yeshu Ben Pandira', or 'Jesus son of Pandira'; to his having been in Egypt where he practiced magic; that he was a Nazarene who was 'hanged', who died young; and to his mother Mary being of royal descent. These descriptions, however, fail to place this Yeshu in an historical context. Rabbinical writings, for instance, 'at no point... refer to his execution as having taken place during the reign of Herod or when Caiaphas was high priest'. He may therefore have been a great teacher who had lived a long time earlier. Some believe that, by the end of the first millennium BC, people like John the Baptist were waiting for the *Second* Coming of Jesus, implying that he had lived in the distant past [Baigent & Leigh, 2001, p160, p162, p240].

How, then, did this religion come about, a religion which infected the Romans, and which has dominated our lives in the West for 1,600 years? What has puzzled historiansis that recognizably Christian communities existed early on in the first millennium AD. A sizeable community was in Rome by the time of Nero between AD 54 and 68 [Osman, 1998, p299] – which, at a time of slower communications, was surprisingly quick for a new religion to have spread from the Middle East. The Roman writer Tacitus refers to Christianity as a 'pernicious superstition' and to Nero accusing the followers of 'Christus' of being responsible for the fires in Rome.31 This reference to 'Christus' further confuses the picture as this is another word for 'anointed one' or 'messiah'. The 'Christus' that Tacitus mentions could well have been Paul, rather than Jesus, as there is evidence of Paul being referred to by that title [Picknett & Prince, 1988, p433]. In a much later Mandaean *Book of John* there is reference to 'Messiah Paulis' as being *'Christ the Roman'*.

The Year Zero Event
So what might have happened? There are constant references in the Essenes' Dead Sea Scrolls to a 'messiah'. The messiah that they were anticipating could have had either a spiritual or political role, because traditionally both priests and kings were anointed. A later allusion in *Hebrews*, the New Testament book, to Jesus having become 'a high priest forever after the order of Melchizedek' emphasizes this confusion between the two roles: the name *Melchizedek* being the combination of 'king' (*melek*) and 'priest' (*zadok*) [*Hebrews* 6:20, *Hebrews* 5:5-6].

There were probably many in first millennial Palestine who hoped that a messianic king figure would replace the puppet Herodian dynasty, liberating the country from the hated Roman oppressors. Certainly groups such as the Zealots had been actively hostile toward the Romans for a long time, organizing revolts against them since the 2nd century BC. Yet regardless of all this turmoil, no single messianic figurehead seems to have emerged, either to rescue a kingdom or to rescue souls and start a new religion. While the Essenes may not be able to provide historical evidence for the existence of a messiah – Jesus or otherwise – connections between this Jewish sect and the early Christian community are undeniable.

In many respects the Essenes had parallels with modern puritan societies like the Amish in America – both groups lived in well-regulated, self-contained communities with a strong work ethic and strict rules of admission. Essenes were fastidious about meals, which were sacred, with prayers and blessings. They were served food quietly in a strict hierarchical order. Before eating they always bathed, wholly immersing themselves [Dupont-Sommer, 1961, p28, p23, p25, p27 p29, p151], changing their clothes in order to be 'pure'. They were especially disdainful of what they called 'seekers after smooth things'. Anyone who wanted to join the community had to spend time as novitiates in order to prove themselves worthy of admission [Allegro, 1981, p29, p30]. They attacked wealth – being in many senses early communalist societies who shared everything in common. On joining, individuals surrendered all their worldly goods, including clothes. In return the community cared for them when they became old and sick, and educated their children.

The direct influence of the Essenes on Christianity is particularly evident in the episcopalian structure of the Church, based on the Essenes' concept of the overseer who had the title 'Cephas'. In the *Book of John*, Jesus specifically says to Simon 'You shall be called Cephas (which means Peter)' [*John* 1:42]. Paul also refers to Cephas in *Galatians 1:18-20* [Baigent & Leigh, 2001, p265]. The prestige of this title is reflected in the Essenes' community rule description of the Cephas as the 'precious cornerstone' [Allegro, 1981, p206, p218].

Furthermore, there are clear echoes of both Christian communion and the Last Supper in the following extract from the Essenes' pre-Christian Rule of the Congregation – as well as interesting references to 'messiah':

And when they are gathered together at the common table to

eat and [to drink] new wine and the communion table is laid out for eating... let no man extend his hand over the first-fruits of bread and wine before the Priest (i.e. the Priestly Messiah); for he will bless the first-fruits of bread and wine... first. Thereafter, the Messiah of Israel (i.e. the lay messiah, the prince) and each... of the Community, each according to his dignity, will stretch his hand over the bread... It is according to this prescription that they shall proceed at every ritual meal at which at least ten men are gathered together [Allegro, 1981, p115].

Nazarenes

How did this confusion between the Essenes and the Christians arise? Part of the answer possibly lies in Matthew's statement about Jesus, namely that 'He shall be called a Nazarene' [*Matthew* 2:23]. Paul also consistently refers to Jesus as a Nazarene [Osman, 1998, p163]. The early Christian father Epiphanius states that 'Early Christians were known as Nazoreans in Judea' [Baigent & Leigh, 2001, pp255-56] and it is a term that continues to be reflected in the modern Arabic for Christian, *Nasrani*. But, as 'Nazarene' could not mean someone from Nazareth, because Nazareth did not exist at the time, it had to have a different meaning.

Far from being a new phenomenon that appeared with the assumed arrival of Christianity, it is possible that the Nazarenes were a well-established group with a long history. They may have been the people that God spoke to Moses about in the Old Testament, called *Nazirites*. These were people who took special vows; they did not drink alcohol (nothing from grapes); and they did not cut their hair until their 'time of separation' was over [*Numbers* 6:1-13]. The clue lies in the common semitic root word, *nsr*, that 'Nazarene', 'Nasrani', 'Nazorean' all share, which has a meaning 'to guard' or 'to protect' [Osman, 1998, p163]. The same root also appears in the Hebrew expression *Nozrei ha-Brit* which translates as 'Keepers of the Covenant' [Baigent & Leigh, 2001, p256], a typical description that the Essenes used about themselves. 'Nazarene' could therefore have been another term for an 'Essene'.

The Essenes regarded themselves as the 'Keepers of the Covenant and the Law', and as 'Men of the New Covenant or New Testament' [Allegro, 1981, p44], 'the Perfect of the Way, the Way of Righteousness'. If the Essenes and the Nazarenes were the same, and the Nazarenes have since been identified as being early Christians, then it is not surprising that there were communities of them in existence before Jesus. That would explain why Christianity appeared to have spread so quickly in the 1st century AD and why early Christians and the Essenes had so much in common – because at one time they were essentially the same people.

Yet for all their similarity to Christianity, the Essenes remained a resolutely Jewish and not Christian sect. What is certain about the Essenes is that they had a strong sense of their own Jewish self-importance. They were most disapproving of other Judaic sects. They criticized the Temple in Jerusalem,

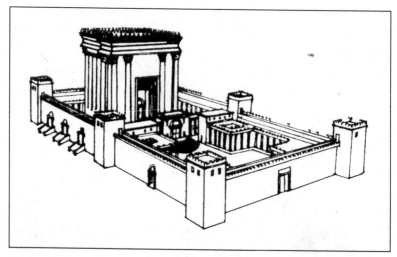

Reconstruction of Herod's Temple

considering it to be incorrectly configured. They thought it too small, and it should have had three courtyards instead of two [Feather, 1999, p219]. They kept Jewish festivals at different times to other Jews, because they followed a solar and not a lunar calendar. Unlike those in Jerusalem, their festivals also did not involve animal sacrifice.[32]

The Essenes and Akhenaten

What gave the Essenes their exaggerated sense of self-right-eousness – and the opinion that everyone else was wrong – was their belief that only they followed the orthodox version of the Jewish faith. Only they were the inheritors of the true tradition, a tradition that came from Moses (or perhaps, more particularly, from Egypt and Akhenaten?). The Essenes may have considered themselves closer to their Egyptian heritage than other Jewish sects because they could have been among those who remained Egypt instead of leaving at the time of the Exodus.

Although, by the 2nd century BC, the Essenes were celebrating Passover, they otherwise participated in only three other Jewish festivals: the Feast of Weeks (Pentecost), the Day of Atonement (Yom Kippur) which they observed with a 'messianic banquet' rather than the usual Jewish animal sacrifice [Osman, 1998, p225], and the Feast of Tabernacles. Missing was the fifth festival of Purim concerning events described in the *Book of Esther* which the Essenes did not possess, presumably because the Essenes were not in Israel at that time. They were also absent when the second Temple was built.

The Essenes could have stronger associations with Egypt because they may have formed the strange Jewish community that existed up to the 5th century BC on Elephantine, an island in the Nile near the southern border of Egypt. This community was identifiably Jewish in spite of clear differences with Jews in Israel. It may have been there since the time of the 18th Dynasty. There is archaeological evidence on the island of temples and other artifacts dating to the time of Tuthmoses III and Amenhotep III, as well as the clear influence of Akhenaten [Feather, 1999, p255].

When the Persians invaded Egypt in 525 BC they found an unusual form of Judaism in which both Yahweh and Astarte were worshipped but there was no awareness of the Oral Laws and no celebration of Passover (not surprisingly). Indeed, in 419 BC Darius II was asked to give the Elephantine island community instructions on the Feast of Passover because they lacked knowledge of it. They also still spoke Aramaic even though Hebrew was the language of Israel from the 8th century BC [Feather, 1999, p254, p256, p253].

Undoubtedly, the Essenes retained several strong cultural links with Egypt. The Dead Sea Scrolls show that, apart from *Isaiah* and the *Book of Enoch*, the Essenes had texts such as *Jubilees* which are generally reckoned to have originated with them [Feather, 1999, pp212-213]. *Jubilees* in particular is more an Egyptian than a Jewish document; not least because the concept of a jubilee relates to the Egyptian pharaonic ritual of the *Heb Sed* [Dupont-Sommer, 1961, p98].

Another powerful connection with their Egyptian past was the Essenes' attitude to the Sun. Akhenaten's radical concept of a monotheistic religion could be simplified as a form of Sun-worship, with the 'Aten' as the abstract deity, the force behind the Sun, and the Essenes continued with the veneration of this force in an obvious manner. Every morning they dedicated ancestral prayers to the east, before sunrise [Dupont-Sommer, 1961, p28] – a practice which gave rise to other names for them, the 'Sons of the Dawn' or the 'Sampsaeans'.[33]

The Copper Scroll

But perhaps most intriguing of all is the direct evidence linking the Essenes with Akhenaten. This evidence exists in one of the Dead Sea Scrolls, the Copper Scroll, which has been the subject of much speculation since it was found in Cave 3 at Qumran in 1952. This scroll has attracted interest not least because it gives precise details, in terms of weight and type, of fabulous treasure including gold and other precious metals, as well as identifying their various hiding places.

The treasure is usually assumed to have come from the Temple in Jerusalem. Writers Michael Baigent and Richard Leigh are typical in their opinion that the scroll was an 'accurate inventory of the Temple of Jerusalem', but that the locations in scroll have been 'rendered meaningless by time, change and the course of two millennia' [Baigent & Leigh, 2001, p50]. Another analyst, a former metallurgist named Robert Feather has, however, a different interpretation.

Feather believes that there are convincing connections between the Dead Sea Scrolls and Egypt. Three of the Dead Sea Scrolls are written using a red ink which would have been unusual in Judea but not in Egypt. Analysis of this ink has shown it to contain a mercury compound which was not present in Israel [Feather, 1999, p230]. Secondly, with regard to the Copper Scroll itself, Feather considers that the choice of copper as a material for a scroll was not normal outside Egypt. A similar copper scroll found at Medinet Habu was also a temple inventory. He thinks that the copper of the Copper Scroll is also of a very pure copper, 'almost identical in chemical composition to the copper being produced in the [Egyptian] 18th Dynasty'.

The content reveals further connections. The scroll gives precise information about what and where the treasure is to be found. It states that 65 gold ingots are to be found in the 'third platform' of the Old House of Tribute, or in the 'carpeted house of Yeshu' [Jesus?]. There is one other mention of 100 bars of pure gold, giving a total of 165 bars. In 1926 a team of archaeologists found a lead weight at Tell el-Armana, the site of Akhenaten's city Akhetaten, which could have been used to standardize the weight of a gold ingot. This lead weight weighed 20.4 grams.

Extraordinarily enough, this same team also found bars of gold weighing a total of 3,375.36 grams in an estate courtyard in another part of the city. This weight of gold is close enough to the total of the bars listed in the Copper Scroll, multiplied by using the lead weight [i.e. 20.4 grams × 165 = 3,366 grams] [Feather, 1999, p172], a coincidence which encourages the belief that the treasure was originally buried at Tell el-Armana.

The scroll also describes the locations of the treasure as being in a city that most closely correspond to the idealized city mentioned in *Ezekiel* in the Old Testament and in *Revelations* in the New Testament [*Ezekiel* 40-48; *Revelations* 21]. The same idealized city appears in six manuscripts of the Dead Sea Scrolls. As a consequence, these manuscripts have been given the title the New Jerusalem Texts, even though there is no mention of Jerusalem in any of these texts and the idealized city plan does not relate to any known layout of Jerusalem at any time in its history.

Feather reports, however, that when a team of researchers compared the New Jerusalem Texts in the 1980s with actual cities in the ancient Near East, they found that the best fit was Akhetaten [Feather, 1999, p146]. Especially convincing was *Ezekiel*'s description of a large river to the west of the temple, correlating to the relationship between the Nile and the Great Temple of Akhetaten, whereas there is no such river in Jerusalem [*Ezekiel* 47:1-12; Feather, 1999, p220].

The Birth of Christianity

If the Essenes developed their beliefs and practices as Jewish offshoots of the Egyptian Atenists (with their Sun-worship), how did Christianity become something more than an 'Essenic' version of Judaism and the original Egyptian Atenist religion? The answer would appear to have involved a spat between Paul and the Essenes. Christianity would not have happened if Paul had not had a falling out with the Essenes.

But we have been denied this explanation because the relationship between Paul and Christianity, as presented in the New Testament, does not tell the full story. While it is scandalous that it has taken so long (more than 50 years) for the transcriptions of the Dead Sea Scrolls to be made public, the Jesuits' reluctance to publish is understandable. After all, they might have to admit that the religion which has given meaning to their lives could be in danger of collapsing.

For a start, it makes no sense that Paul as Saul of Tarsus would have been sent on a mission on behalf of the Temple in Jerusalem to Syria, when the Jewish Temple had no jurisdiction there [Baigent & Leigh, 2001, p221]. The reason for sending him to Damascus becomes clear only when it is realized that 'Damascus' was a code name for Qumran, the Essenes' site next to the Dead Sea. It was a name commonly used by the Essenes – as evidenced in the title of one of the Dead Sea texts, the *Damascus Document*, ten copies of which were found at Qumran [Baigent & Leigh, 2001, p219].[34] The problem for the Jesuits is that, although the Dead Sea Scrolls make no reference to Jesus, and there is little in the scrolls that corresponds to the canonical gospels, they still describe certain events as having taken place.

The scrolls repeatedly refer to the Righteous Teacher who was killed by the Wicked Priest, and to their faith being betrayed by the Liar. Since the Jesuits cannot deny these references, what they have ingeniously decided is that this story must have taken place at some time in an earlier period, in the 2nd century BC during the time of the Maccabeans, in order to avoid any confusion with the possibility that these three people and these events might have occurred around the time of Jesus. But they have taken this decision without any proof that such was the case [Baigent & Leigh, 2001, p226].

There is, however, an alternative interpretation of the Scrolls which makes sense in the context of the early 1st century AD. This is a context in which Palestine had been hostile toward the

Romans for some time and was particularly unhappy with direct Roman rule since 63 BC [Mack, 1993, pp54-55]. The Essenes, far from being the peace-loving people that they are often portrayed, had also organized their site at Qumran as a fortress as much as a monastery [Baigent & Leigh, 2001, pp229-30].

It had thick defensive walls which were strengthened with towers and a rampart when rebuilding began after 4 BC, and a forge that could have been used for making weapons. They were also violently against the Romans, who they referred to as the *kittim* in their texts. The Essenes' War Scroll has great detail of battles to be fought against the *kittim* and their king – implying that the *kittim* belonged to the Roman imperial age after the end of the Republics in 27 BC [Baigent & Leigh, 2001, p237, p214].

What concerned the Temple priesthood in Jerusalem was the existence of a rebellious community at Qumran, so openly opposed to the Romans and so close to the city. They clearly feared that the proximity of such a community would draw the anger of the Romans to Jerusalem, thus jeopardizing the safety of the Temple itself – hence the mission of Saul to 'Damascus' to confront the Essenes and reduce the threat. He failed in his mission and in AD 66 the priests' fear was realized: the Romans destroyed the Temple as well as the Qumran community.

By that time, Saul had had his damascene moment and, instead of attacking the Essenes, he had decided to join them. The Bible mentions that he spent three years in the wilderness in Arabia – most likely as a novitiate in one of their communities there, perhaps even at Petra in Jordan, before returning to 'Damascus' (Qumran) as a full member of the sect [*Galatians* 1:17]. Having changed his name to Paul and having been accepted into the Essenes, he later decided to leave them to set up his own group.

The Essenes were not happy about this decision and were forthright in their description of him as a 'liar', as someone who has betrayed their community. The Damascus Document repeatedly denounces 'the liar', those who 'enter the New Covenant in the land of Damascus, and who again betray it and depart', and those who 'deserted to the Liar' [Baigent & Leigh, 2001, pp223-224, p286, p220]. The Dead Sea Scrolls are thus actually descriptions of the row that the Essenes had with Paul, the Liar – a charge which Paul vigorously denied in his various letters in the New Testament.

This event occurred at about the same time that the Essenes' Righteous Teacher – who could have been James the Just (known to us as a brother of Jesus) – was killed by the Wicked Priest, who could have been Ananas,[35] the high priest in charge of the Temple in Jerusalem, who had James thrown from the roof of the Temple – thus placing the story in AD 62 rather than earlier [Baigent & Leigh, 2001, p279].

A New Religion

Where Paul differed with the Essenes was on the matter of the Law [Baigent & Leigh, 2001, p269]. The Essenes feared that Paul would lead people away from the Law of Moses, whereas Paul wanted to create communities based on faith rather than Law, communities that were accessible to all, Jews and Gentiles – hence the discussion about accepting the uncircumcised into the Church [*Galatians* 2:7-21]. He wanted to simplify admission into the community without the need for elaborate initiation. But even from the beginning, there were tensions within Paul's new community.

On the one hand, there were the Gnostics, one of whom was probably Paul himself – people who were interested in the ancient wisdom teachings that came from Egypt, which eventually reappeared in what is now known as the *Hermetica*. On the other hand, there were groups within the early Church for whom it mattered greatly that their religion was founded on more solid 'facts'. They could be characterized as 'literalists', perhaps represented by the person of Peter, the Cephas or 'cornerstone'.

In essence, the Gnostics' explanation for the existence of wickedness in the world was, as John Allegro puts it, that 'Light had become mixed with darkness, spirit with matter. All living beings carried within them the spark of divine light, but its imprisonment in the flesh meant they could no longer communicate with its source...' [Allegro, 1981, p105]. In holding this belief, the Gnostics were following the Essenes who, as Josephus, the Roman historian, points out, also believed in the immortal soul trapped in a physical body and, says Josephus, their views were like the Greeks and compared with the principles of the Pythagorean schools [Allegro, 1981, p245]. Whereas Plato or Pythagoras acknowledged the 'spark of divine light' and emphasized the study of mathematics or music as the means for reconnecting with the eternal harmonies [Frend, 1984, p170], for Gnostics the teacher was the essential 'mediator of [divine] knowledge, as it was only those with gnosis – transcendental knowledge – who would be saved and able to return to fullness, the *Pleroma*, from which they emanated' [Allegro, 1981, pp105-06].

On that basis, the importance of Jesus to the Gnostics lay in what he said, not who or when he was. For that reason their texts contain no reference to the historical person of Jesus. The Egyptian *Therapeutae*, who buried some of this literature in about AD 300 at Nag Hammadi in Egypt, were the 'Gnostic successors of the Palestinian Essenes', carrying on many of their beliefs and practices – even though they claimed Isis as the 'Mother of God', until the cult of the Virgin Mary took over [Allegro, 1981, pp152]. The *Gospel of St Thomas*, copies of which were rediscovered at Nag Hammadi in 1962, is typical of this wisdom literature in the Egyptian tradition.

Some biblical researchers believe that they can identify remnants of a lost gospel embedded in *Matthew* and *Luke* [Mack, 1993, pp1-3, p20], on which *Thomas* may have been based

[Osman, 1998, p233]. These researchers suggest that this lost gospel was a book of Jesus' sayings, filling the gap between *Mark* and the other two canonical gospels, *Matthew* and *Luke*. They believe that this gospel belonged to a group they call 'the people of Q', who were not necessarily even Christian.

The people of Q, if they existed, had much in common with the Cynics, a loose grouping of people who drifted throughout the Classical world in the 1st century AD with a particular philosophical approach to life. The aphoristic style of Q is close to the Cynics' way of making pointed comments on human behavior. Also, like the Cynics, the people of Q were expected to live constantly on the move, parted from their families and reliant on charity from the people they encountered. These people were not interested in Jesus as the Messiah or in a cult of Christ who died to save us, but in Jesus as someone who taught a way of living [Mack, 1993, p46, p4].

The Appeal of the Literal Truth

Although the beliefs of the Essenes and their Gnostic successors may have been the authentic legacy of the original Egyptian Atenism, theirs was not an interpretation of Christianity that has prevailed. The literalists, who created the myth of St Peter as the first Bishop of Rome, won the day. These were people who wanted the reassurance of believing in an historical figure who had lived in a timeframe to which they could relate – such as changing the manner of Jesus' death to crucifixion rather than by hanging, to give the story a Roman context [Osman, 1998, p164].

They clearly had a great need for certainty. They wanted a canon of orthodox teachings, fixed around the comforting biographical details of a spurious life while, at the same time, placing their trust in a highly hierarchical organizational structure inherited from the Essenes, in the hope of achieving eternal salvation. It was for these literalists that the canonical gospels were chosen, written in the style of eye-witness accounts, even though they could not have been written during Jesus' supposed lifetime – the earliest Christian document was possibly Paul's *First Letter to Thessalonians*, thought to have been written in AD 51-52 [Brox, 1994, p132]. Scholars are fairly certain that *Mark*, the first of the canonical gospels, was written about AD 70 – not least as *Mark* makes an oblique reference to the destruction of the Temple which occurred in AD 66 [*Mark* 13:1-2] – followed by the others at ten year intervals.

It was also the literalists who successfully promoted Peter over the more Gnostic Paul in order to create an apostolic line of authority for the head of the church in Rome, so increasing the influence of Rome in the power struggle between the other centers of Christianity at Alexandria and Antioch. It was they who began to hold councils by the end of the 2nd century AD to fix the date of Easter; they who, by AD 200, rejected Gnosticism as a heresy linked to Greek philosophy; and it was they who wanted to establish a single recognizable Rule of Faith that could be circulated to all main centers of Christianity. Above all, it was the literalists who continued to develop the Church's hierarchical structure with the result that, by the mid-3rd century AD, there were many different minor orders of 'presbyters, deacons, subdeacons, acolytes, exorcists, readers, doorkeepers' and so on. By this time the bishops had also come to rely on tithes for their monthly salaries [Frend, 1984, pp282-284, pp404-405].

Against this background the Romans began to adopt the new religion during the 4th century AD. Given the similarities between the Essenes and the early Christians, it is understandable that we should be confused distinguishing one from another. But it is the differences between the Gnostics and the literalists that have shaped what we know today as Christianity.

Understandably in such a climate, the Gnostics buried their texts at Nag Hammadi in ad 360-67 [Frend, 1984, p577], including part of the *Hermetica* and two sections from the Aesclepius [Merkel & Debus, 1988, p49], and the literalists came to dominate.

With God in charge, the West truly began the Dark Ages and even European cities began to disappear, including those that the Romans had built.

Chapter 17

How the West lost Civilization

The 4th century AD was a fatal century for civilization in the West. Within a hundred years the Western world changed irrevocably as the Romans went from the persecution of Christians to the adoption of their religion.. At the beginning of the century, in October 312 AD, Constantine, the Roman general, legendarily saw the *chi rho* symbol of the Christians in the clouds before the battle of Milvian Bridge on the Tiber, where he defeated Maximian and Maxentius and became emperor. It was Constantine's dynasty who closed the gruesome 'Era of the Martyrs' – the savage killing of Christians that had happened on and off since the time of Nero [Frend, 1984, p481] – thus making it possible for Romans to become Christian. But whereas Constantine had promoted a freedom of worship that included Christianity in his open-

Temple of Isis on the island of Philae, Egypt

minded Edict of Milan in AD 313 [Frend, 1984, p483], by the end of the 4th century, Christianity was the only religion allowed.[36]

In AD 380 Emperor Theodosius I issued an edict insisting that all citizens of the empire must become Christians – effectively banning paganism. The way was then open for bands of fanatical monks to destroy all vestiges of paganism. These were the people who killed the priests at Philae in Southern Egypt, who destroyed the site of the Eleusian mysteries in Greece in AD 396 [Freke & Gandy, 2000, p22] and burned down the famous library at Alexandria in AD 415, brutally murdering Hypatia, its last priestess.

By the end of the century the Romano-Christians had symbolically and literally cut us off from the ancient past. When they massacred the last Egyptian priests at the Temple of Isis on the island of Philae in AD 394, all knowledge of hieroglyphs also disappeared – around this time the last hieroglyphic inscription was engraved on the island. As a result, one means of understanding Egyptian civilization and all of its wisdom and knowledge was soon forgotten, only rediscovered in the 19th century thanks to the Rosetta Stone [Merkel & Debus, 1988, p31].

What also changed in the 4th century AD was that, because of Emperor Constantine, Christianity in effect became a romanized religion, the Roman Catholic Church. The process was symbiotic. The Romans adapted the new religion to the extent that a lot of what we understand by Christianity dates from then. Meanwhile, the Church inserted itself into the institutional framework of the Roman Empire, successfully building a parallel state from within and which in the end, in a sense, took over the empire.

Christianity as Progress
On the one hand, one might well ask why it should matter that the Romans enabled Christianity to dominate our lives, if we

have now progressed. Isn't belief in One God more 'advanced' – even if a few mistakes were committed along the way – and isn't polytheism a sign of a backward society? Surely we wouldn't be the 'civilized' people we are today without the combined influence of the Romans and the early Christians? Our success in creating a modern society based on science and technology must be due in part to the brave individuals of the past who helped to rid the world of superstitious pagan nonsense.

On the other hand, what happened 1,600 years ago matters because we are still living with the consequences of the decision of the emperor Constantine to adopt Christianity. It matters because what happened in the 4th century AD affected our understanding of 'truth'. Fundamentally, Christianity, under the influence of the Romans, altered the philosophical basis for civilization in the West. By substituting the complex pantheism of the old beliefs for the simpler hierarchy of the Holy Trinity, with an all-powerful God the Father figure at the head of it, the Church effectively began a process of 'unbundling' and disengaging the holistic nature of ancient religious thought that had been central to civilization.

Pantheism had, in any case, always acknowledged a superior god – often some form of a Sun-god such as the Greek Apollo, whose name meant 'the not-many', or 'the One' – even though it embraced a variety of deities. Celsus, writing in the 2nd century AD, pointed out Christianity's similarities with other doctrines, stating that 'It matters not a bit what one calls the supreme God – or whether one uses Greek names, Indian names or the names formerly used by the Egyptians'. But in the old religion there was no attempt to use the superior deity to monopolize 'truth'. In Egypt, each temple had its own special deity, and no single temple dominated the country or its religion – until, significantly, Akhenaten closed all temples in favor of the Aten.

The Catholic Church reinforced its beliefs with actions, as demonstrated in the choice of location for new churches. The

Church deliberately chose the old pagan religious sites for its places of worship, turning the enclosed areas around them (the *temenos*), previously used for ritual dances, into churchyards and places of burial [Pennick, 1996, p115]. In taking over these sites Christianity purposefully destroyed the respect that the old religion had for the forces of nature – the derogatory use of the word *pagan* itself meant 'country dweller', or even 'bumpkin'. Thus the Church transformed the Greek god of nature, Pan, into an emblem of the devil with his cloven hooves and his horns. The Green Man, with his mouth spouting leaves and branches, carvings of which one can occasionally find hidden away in medieval churches, became a subversive image. The Church wanted to disconnect us from nature.

But what Christians misunderstood was that pantheism explained the world by allocating specific attributes or principles to different deities and forces within nature. Egypt was of course the example *par excellence* of this practice. Egyptian pantheistic religion wasn't necessarily superstitious: it was just easier to comprehend different aspects of the One God if they were distributed among lesser deities who could contribute a many-layered understanding of the world and its mechanisms, both profane and profound.

The point of a god that dies and is resurrected at Easter, for instance, is that he is a fertility symbol of life returning to the land after the winter. An essential aspect of the Egyptian Osiris was his role as a vegetative deity – the festival of Khoiak 'began with a ploughing and sowing ceremony, which symbolically corresponded to the death, dismemberment and interment of Osiris in the earth' [Naydler, 2005, p244].

It is only now that we are rediscovering the importance of this ancient attitude – for example, former NASA scientist James Lovelock's symbolic use of the Greek Earth goddess *Gaia* as a means of restoring our sympathy for the Earth. Celsus, the 2nd century AD Greek commentator, understood this point. He knew that 'the religion of Egypt transcended the worship of the irrational beasts'. He realized that 'the animals were symbols of invisible ideas and not objects of worship in themselves' [Hoffmann, 1987, p71].

Authoritative Truth

Paradoxically, the Church's attack on the hierarchical pantheism of deities had the effect of reinforcing a mundane world of hierarchical ecclesiasts instead. As early as AD 115, Ignatius of Antioch had tried to justify the Church's structure because it reflected 'heavenly order' [Brox, 1994, p80]. After AD 185, when the principle of apostolic succession became established, Bishop Irenaeus of Lyons claimed that the truth was exclusive to bishops because 'they alone took over the truth from the apostles and preserved it'.[37]

Because the Church wanted to monopolize 'truth' – only the 'Word of God' was correct – it then followed that only bishops were able to interpret it, and then only the supreme bishop, the Bishop of Rome, the Pope.[38] Debate within the Church was not encouraged, as its doctrines were based on faith, not reason. The problem with this kind of approach is that it was entirely open to abuse and arbitrary interpretations which could not be tested or reviewed: an action or a belief must be acceptable if God's representative says so. Bishops, who had begun as administrators of the community, had by the 3rd century AD become immensely powerful.

Such was the power of the Church that if a bishop excommunicated someone the consequences were serious. As well as social disgrace, the 'public confession of sins before the community could cost people their professions'. Even emperors were not immune to the threat of excommunication. There was much discussion about the relationship between the emperor and the Church as there was no theory in Christianity with regard to the position of the emperor. In the end it was decided that the emperor was 'in the Church' and not 'over' the Church, so as to clarify the Pope's position as being overall 'father' [Brox, 1994, p114, p53, p60].

The Pope as Emperor

It was under the Constantine Dynasty that the parallel Catholic state started to build, developing its own fiscal and legal structures which were recognized independently within the empire. From AD 318, Constantine gave bishops' courts the same rights as a secular magistrate's court, and bishops could sit in judgment in trials involving Christians [Brox, 1994, p64]. The jurisdiction of the church and secular courts became equally valid, and a person could choose which court should hear his case [Frend, 1984, p488]. In addition, not only did Constantine exempt churches from local taxes but, in AD 349, his son allowed them a share of local levies as well [Frend, 1984, p537, p487]. The Church became a 'vast, privileged corporation'.

Just as significant were the alterations to church ritual and doctrine in the 4th century. One change that we still abide by, without realizing it, was the decision of Pope Liberius in AD 353 to make the date of Mithras' birth, 25th December, the birth date of Jesus instead. Until then the Church had had a major festival on 6th January for the baptism of Christ rather than his nativity [Gilbert, 1996, p61]. Indeed, 6th January still remains an important festival in Orthodox countries today.

The Eucharist also lost the meaning that it had had as a community meal. It became more 'the performance of sacred rites and... a public act of worship'. Changes to Christian doctrine have been equally long lasting. It was in the 4th century that the canon of holy books was defined, excluding all traces as far as possible of heretical Gnosticism [Brox, 1994, p130]. Particularly important was Emperor Constantine's decision to convene the Church's Council of Nicaea in AD 325, where the Church Fathers agreed on the wording of a creed that remains almost unchanged today.

This close relationship between Church and State meant that, even after the fall of the Roman Empire in the early 5th century AD, in a sense little changed. Through the institution of the papacy, the Romans retained psychological and spiritual connections with and control of former imperial citizens. Pope Leo I (AD 440-461) was more than happy to fill the vacuum that formed in Italy in the 5th century AD due to the lack of an emperor in the West, becoming 'a powerful figure with a corresponding court ceremonial' [Brox, 1994, p91]. As imperial Rome weakened and temporal power shifted to the Germanic-speaking Holy Roman Emperor, with whom it stayed for many centuries, the Roman papacy gained in strength. The rivalry that existed for several centuries between two of the centers of Christianity – Alexandria and the Near East (Antioch to begin with, and later Constantinople) – further enabled the Roman papacy to dominate the Christian world [Brox, 1994, p131, pp86-88].

From the 5th century on, the Pope thus became a substitute for the Roman emperor, with the Church acquiring many of the

A Byzantine apse

externalities of the empire. In the place of the Senate, the Pope was notionally elected by the cardinals. Instead of an imperial palace, the Pope took centre-stage on a throne with a baldachino in an ecclesiastical basilica built with an imperial-style apse under a triumphal arch [Brox, 1994, p65]. As well as the throne, he had other imperial trappings acknowledging his social status [Chadwick, 1993, p164; Brox, 1994, p64].[39] Meanwhile, at the other extreme, people voluntarily enslaved themselves in monastic orders to work for the Church – Rome having always had an unending supply of slaves from everywhere in the empire. The Roman church also continued to receive booty – only this time it was in the name of God.

The status of the Church even affected the fabric of civilization. As the Western Roman Empire unraveled in the 5th and 6th centuries, cities became physically and psychological subordinated to the Church, with ecclesiastical buildings, churches and abbeys becoming architecturally significant. Ironically, the ideal of the city survived with Augustine as the 'spiritualization of the memory of Jerusalem and the concept of Roman *civitas* in his City of God' – a situation which did not change until the late 1300s, the early Renaissance, when in Italy there arose a renewed interest in urban design [Merkel & Debus, 1988, p405].

Ignorance and the End of Divination

But it was not just our thinking about the world that was affected. The attack on paganism resulted in the loss of useful, practical knowledge regarding the physical world. A common characteristic of being 'ruled' by God seems to be that knowledge bases are destroyed – a phenomenon not just restricted to the 4th century AD. One can see the same rejection of scientific knowledge today in favor of a narrow theological interpretation in modern theocracies in the Middle East or educational establishments in the West run by 'creationists'.

Throughout the 5th and 6th centuries Roman emperors intermittently continued the attack on pagan knowledge on behalf of the Church. The destruction of the famous library at Alexandria in AD 415 was not an isolated event. One hundred years after the transfer of scholars from Alexandria to Athens, Emperor Justinian I closed the university in Athens in AD 529, simply because it was pagan [Brox, 1994, p51]. With the demise of imperial Rome, the Roman Catholic Church carried on the assault.

Having the birth of Christ as a *tabula rasa*, a clean slate, left the Church in a quandary. It effectively rendered dubious any knowledge acquired before the Year Zero because such knowledge lacked authority, leading to peculiar discussions about the Greeks – especially 'saint' Plato – not least because the Greeks were not necessarily seen to be 'bad' pagans, having learned a lot from the Egyptians since the middle of the first millennium BC.

To begin with, we were allowed to know what the Greeks knew only if it found favor with the early Christian fathers. Thus, Ptolemy and Aristotle determine our early understanding of science, but Plato had to wait for St Augustine before he was redeemed. It was largely due to the Greeks' love of mathematical and mechanical order that Newton and others later developed an incomplete interpretation of the universe as essentially a giant clockwork operating as a solid body according to certain fundamental laws.

Thanks to the deliberate destruction of knowledge, centuries later we had to painfully *re*discover two key facts about the universe: that the Earth is round and that it orbits the Sun. It was the Greek pharaoh Ptolemy I who had developed a cosmology with the Sun going round the Earth, which was absorbed into Christian doctrine. The pre-Ptolemaic ancients were well aware that the universe was heliocentric and that the Earth was a globe [Clark, 2000, p438; Hancock, 1993, p335]. The *Hermetica* describes Ra the Sun god as being 'placed in the centre and wearing the Cosmos like a wreath around him' [Freke & Gandy, 1997, p72].

Most important, the Church did not want any competition for access to God. As a result, what disappeared, along with the destruction of the libraries and the persecution of the pagan priesthoods, was any metaphysical method of accessing knowledge. Any form of enquiry, such as the casting of horoscopes or other methods of divination, became anathema to the Church. Only the Church could intercede with the Divine, and anyone else who attempted to do so was in league with the Devil.

In early Greek astrology the original word for 'sign' or 'omen' was *daemonion* (according to Socrates, *daemons* communicated through signs and symbols), but what the Church did was to change the idea of the 'good daemon' [Cornelius, 1994, p191, p129] into something evil, and the word 'demon' acquired its negative connotations. St Augustine was one of those who endorsed the persecution of pagans with his condemnation of Hermes (Thoth) as 'having inspired the demonic cult' [Merkel & Debus, 1988, p125].

This demonizing of the metaphysical has had a damaging effect on the progress of science and on us. With metaphysics out of the question, because knowledge thus gained was condemned as the work of the Devil, and with the Greeks accepted reluctantly because they pre-dated Christ, it was perhaps inevitable that scientific methodology would eventually develop in the direction that it has. Eventually the *only* acceptable route to *re*discovering information would be by stripping out the subjective from the objective, making the entire process transparent, demonstrable, inherently logical as well as repeatable. Even so, many, such as Galileo, as late as the Renaissance, lost their lives for promoting scientific facts that we now take for granted.

Death and Reincarnation

Arguably, we might not have lost our way, facing enormous problems today if we had not developed such separatist thinking, which clearly predates Descartes. One might say that the roots of our modern-day collective mental illness are to be found in the decisions of 4th century Rome. The romanized Christians reinforced this separation of the spiritual from the secular, of the psyche from the physical. Nowhere is this disconnection more clearly illustrated than in the change in beliefs about death.

We in the West are denied the possibility of belief in death as part of a cycle because of Christian doctrine, and so for us death is a hopeless experience implying permanent separation. As a result, survivors of traumatic events where others have died are sometimes left with terrible feelings of guilt, and the bereaved can feel that life is pointless. They might, however, feel differently if they could believe that, although the flesh dies, the soul lives on and is then liable to incarnation in another body. In the old religion, death was understood as part of the cycle of life. In the words of the *Hermetica*,

> Everything on Earth must be destroyed, for without destruction nothing can be created. The new comes out of the old. Every birth of living flesh, like every growth of crop from seed, will be followed by destruction. But from decay comes renewal [Freke & Gandy, 1997].

It was generally thought in Classical times that the belief in reincarnation originated with Pythagoras in the middle of the last millennium BC [Berresford Ellis, 1990, p89]. But the same beliefs existed independently of the Greeks, among the Celts and the Essenes.[40] For us, Pythagoras is mainly associated with the 'power of number' and the 'mysteries of music'. He is likely, however, to have developed his beliefs on the cosmos and the transmigration of souls from his time in the Egyptian temples. While in Egypt Pythagoras would also have known that the Egyptians had a concept of a Hall of Judgment of the dead soul – a concept which has passed directly into Christian iconography, together with the same scales for the weighing of the dead.

An illustration of this concept can be found on the walls of a 3rd century BC temple dedicated to the goddess *Ma'at*, built by the Greek pharaoh Ptolemy IV at Deir el Medina opposite Luxor and Karnak [Hancock & Faiia, 1998, pp68-72]. In the Egyptian

Egyptian Hall of Judgment

version, the heart, the *Ab*, is weighed against the feather of the goddess *Ma'at*, representing truth, with Osiris sitting in judgment rather than St Peter.

Before judgment, the soul has to pass in front of 42 gods representing the negative confessions, and to each god in turn the soul would have denied committing any wicked acts, such as adultery, murder, theft or polluting the Nile. Then depending on whether or not the heart has weighed successfully against the feather of *Ma'at*, Osiris would send the soul either to join other spirit beings in eternity or to be consumed by *Apep* (also known as Apophis or Ammit, the part-hippopotamus, part-crocodile monster or part-lion) who devoured the souls of the dead in order to send them back into reincarnation [Clark, 2000, p95].

Punishment and the 'True' Cross

The Christian Church may have absorbed elements of the Egyptian Hall of Judgment – St Paul in judgment with his cross keys echoing Osiris' crossed arms, and the same weighing scales for the soul of the dead. But there was no suggestion of reincarnation: the hippo-croc *Apep* was 'airbrushed' out of the process. It did not suit the early Church to have any hint that death was part of a process and that a person automatically had another chance at a mortal incarnation.[41] Instead, the Church promoted the view that 'anything that is sinful brings about a fatal separation from God' [Brox, 1994, p111].

There had been a noticeable shift in Christian theology in the 3rd century AD toward a doctrine of Purgatory, with Origen writing that the 'notion of eternal punishment, *while not actually true*, serves a useful function, since it encourages simpler Christians, still motivated by fear, not to be complacent'. In Origen's view, sinners 'must taste something bitter, so that, having been corrected, [they] may be saved' [Trigg, 1998, p31]. In the preface to his work *Peri Archon*, written c.229 AD, he made clear that 'the soul departs this world... either to inherit eternal life and blessedness... or to be delivered to eternal fire and tortures'. Even in the 2nd century AD, Celsus was aware that the Christians had 'concocted an absolutely offensive doctrine of everlasting punishment and rewards' as a means of persuading people to convert [Hoffmann, 1987, p70].

Concomitant with this shift toward a doctrine of punishment was a change which occurred from the 4th century AD with the increasing use of the Roman cross in the practice of Christianity. The Roman cross, in symbolizing the death of Jesus by crucifixion, represents suffering and punishment more than any other form of cross. Its widespread use in Rome went with the changes in the Eucharist which likewise became 'a recollection of the sacrifice of the Cross' [Brox, 1994, p104].

Until the 4th century AD the cross had had an entirely different meaning and was not a symbol particularly associated with Christianity. Some believe that St Paul used the Egyptian *ankh*, placing greater emphasis on resurrection and eternal life rather than suffering and death [Osman, 1998, pp254-5]. In Egyptian religion the *ankh* had powerful associations with life and the 'endowment of the life force' [Clark, 2000, p14]. Here the attachment of a circle, representing everlasting time, to the arms of a cross signifies manifestation in the physical realm, the cross marking time and place. The Celtic Church, which retained more aspects of the old religion than the Roman Church, continued with this meaning, among others, in its use of a sunwheel on top of a cross, a symbol that pre-dates Christianity by a thousand years [Pennick, 1996, p47]. The equal-armed cross could also be found throughout the ancient world long before Christianity around the necks of ancient kings in Mesopotamia [Baigent, 1994, p4].

Looking Backwards for a Way Forwards

It has taken us a long time to recover from the combined setbacks of the romanized Christians. In particular, the excising of the metaphysical from the physical, of the subjective from the objective, has left us with a distorted understanding of what civilization means. We have continued with the Graeco-Roman philosophical preference for order rather than chaos, reason rather than emotion, ratiocination rather than intuition – in other words, what we regard as a scientific approach.

Now, as we face likely environmental catastrophe and other major disasters, we can hopefully learn from the past about a holistic concept of civilization, based on truth and free will. Perhaps by taking a less conventional approach to the past, and ridding ourselves of inherited Judaeo-Christian triumphalism, can we go beyond the Greeks and the Romans and re-connect with another pattern of development.

We do not, however, need to go back as far as the 4th century AD in order to pick up the threads, as there have been two other major attempts in the last 1,600 years to rediscover the secrets of the ancient past, and so to break the hold that the Church had on our knowledge bases. The first attempt, at the time of the Crusades, was only partially successful. The second, however, had a greater impact on our thinking and even became known as the time of rebirth – the Renaissance.

Chapter 18

The Recovery Process

The romanized Christians nearly succeeded in creating permanent darkness. With the Egyptian priesthood destroyed and any understanding of hieroglyphs gone after AD 400, that could have been the end of civilization. In the West we retreated into a primitive worldview shaped by the Catholic Church. Fortunately for us much ancient knowledge was preserved within the Islamic world, which benefited culturally from the implementation of the *ur*-concept of civilization. It is only with the efforts of brave individuals that we began to re-connect, thanks to the Islamic world, with some of the original concepts of civilization.

At the risk of oversimplifying, the West owes much to two men who, although they lived 300 years apart, were instrumental in weakening the intellectual grip of the Church, so helping to reverse our decline. But whereas the first of them, St Bernard of Clairvaux, *inadvertently* helped a process of change in the first half of the 12th century, the second, Cosimo de' Medici, *actively* sought change in the mid-1400s. Cosimo de' Medici was ultimately more successful in the 15th century – not least because at the time of the Renaissance there was a more general challenge to the Church.

St Bernard, however, had no intention of threatening the *status quo*. He was, after all, a great man of the Church, in close touch with the Pope. Nevertheless, it was through his personal involvement with the Knights Templar that he unwittingly played a pivotal role in the 12th century in recovering lost knowledge.

Islam and Knowledge

Ironically, a significant feature of both periods – the 12th and 15th centuries – was the role of Islam in the Middle East. Because of the preservation within Islam of ancient manuscripts with 'civilized' knowledge, we were eventually rescued from the ignorance to which the Church had condemned us. In the 12th century the Templars became the link with this 'civilized' knowledge. In the 15th century it was manuscripts – known as the *Corpus Hermeticum* or the *Hermetica* – which became available in the West.

Northern Syria – and Harran in particular (the biblical city of Abraham's father Terah) – was also critical to this process of rediscovery, as it was to here that refugees had fled from Egypt from the 2nd century AD onward, together with their ancient texts. Prominent among them were the Sabians, the remnants of ancient astrologers, and the *magi*, who left Egypt in order to escape the Romans. Places like Harran remained havens of safety, even when Islam became the new religion in the Middle East in the 7th century AD.

The Islamic world absorbed knowledge from the *Hermetica*: copies of the *Hermetica* existing in Baghdad as well as Cordoba, where new levels of free-thinking and 'independent reasoning' stimulated Islamic culture's florescence. As a result, from agriculture to architecture, there was virtually no aspect of the ancient concept of civilization that the Muslim world did not know about, from southern Spain to Sicily, to the eastern Mediterranean, as well as in the Middle East itself.[42] It was because of the Arabs that we recovered mathematical knowledge – even the word 'algebra' is from an Arabic term, *al-jabr*, meaning 'the reunion of broken parts' [*Concise Oxford Dictionary*]. We also acquired from them the concept of zero, which had not existed in Roman arithmetic.[43]

Geometry and the Gothic

We may find it bizarre that geometry, which we take for granted when learning mathematics, was once part of the 'civilized' knowledge that was lost to us during the European Dark Ages. But what the Templars rediscovered in the 12th century was the geometry of Euclid, Pythagoras and ultimately of the Egyptians.

Proof of their discoveries lies in the contrast between the Romanesque and Gothic styles of French ecclesiastical building – the latter being a 'new' style that suddenly began to appear in France in the 12th century. The difference between the two styles can be explained by the change from the use of a ruler to a compass in creating the design [Strachan, 2003, p49]. As a consequence, new Gothic churches were no longer based on whole number ratios in arithmetic but instead on the irrational ratios of geometry (that is, the square roots of 1, 2, 3 and 5). The French were inspired to make this change only after they learned about Islamic religious building in the Middle East.

Building mosques with irrational number ratios appealed to Muslims because such ratios reflected their belief that God was immeasurable and therefore unknowable. Geometry thus underpinned the spacing of columns, the typical ogival arches and the *mihrab*, the niche in every mosque that indicates the direction of Mecca. This emphasis on the spiritual appeal of measurement was also echoed in St Bernard's famous statement that God was 'length, width, height and depth' [Hancock, 1993, p306].

We can be fairly certain that Islamic architecture was the primary influence on French Gothic design for a number of reasons. First, there is the superficial similarity of the tall, elegant columns and arches. The three-pointed arch is particularly redolent of Islamic architecture, visible in the design of the west window of every Gothic country church in England. Second, those who had been on the First Crusade in 1097 had passed through western Syria, where the Seljuk Turks had recently built new mosques with the same pointed arches. Third, perhaps more importantly, the new Gothic architecture revealed an under-standing of sacred geometry – albeit applied in an Islamic manner – that had not been present in Europe but was common-place in the Middle East.

The square roots of whole numbers were part of the same sacred geometry as another irrational ratio which came to the West, known as the Golden Section (ϕ).[44] The Golden Section is attributed to a Tuscan mathematician called Fibonacci who lived about 1200 AD, but in fact it was as much a part of ancient knowledge as Pi (π), an irrational ratio that we associate with Pythagoras. As writer Rosemary Clark points out, when π and ϕ are put together they 'evoke harmonic progression, combining the properties of the circle's continuity with the square's incre-mental growth. This phenomenon is termed 'gnomonic expansion' and is a geometric principle consistently found in the architecture of the Egyptian temple' [Clark, 2000, pp200-01]. Analysis of Chartres cathedral reveals the same geometric patterns.

The advantage of using geometry was that medieval archi-tects and stonemasons were able to build their Gothic master-pieces to incredible heights without worrying unduly about the structural engineering of their buildings. As Prof Jacques Heyman, Emeritus Professor of Engineering at Cambridge University, explains,

The problem of the design of masonry is essentially one of geometry. The calculation of stress is of secondary interest; it is the shape of the structure that governs its stability. All surviving ancient and medieval writings on building are precisely concerned with geometry; from the viewpoint of the modern structural engineer, the ancient and medieval rules were essentially correct [Heyman, 1995, p141].

Gothic Influence on the French

It could be argued that the 12th century Islamic influence on France came from the south, either from Moorish Spain or Sicily. Certainly, the Normans encountered Islam in 1063 when they went to Sicily, defeating the Muslims there by 1091, after invading Italy as early as 1017 [Baigent & Leigh, 1997, p83]. But they do not seem to have transferred much knowledge back to France. Spain had also been subject to Islam since the 8th century AD when the Arabs conquered most of Iberia from Morocco. Spain benefited from Arabic knowledge, experiencing an enormous flourishing in culture, with Cordoba becoming the Islamic capital of Spain and a great centre of learning. Yet minimal influence seems to have crept north of the Pyrenees at that time.

If anything, French aristocrats north of the Pyrenees – in the Languedoc and Roussillon – were openly hostile to the Islamic Spanish. In response to demands of the Pope in 1064, they finally launched the *Reconquista* at the end of the 11th century, recapturing Toledo in 1085 [Baigent & Leigh, 1997, p72], to reclaim Spain and protect the shrine of Santiago de Compostela for Christian pilgrims. The French south only later became receptive to Islamic Spain, after ideas from the Middle East arrived in northern France, mainly Burgundy.

In addition, it was later, from the middle of the 12th century onward, that the south of France witnessed the Cathar 'heresy'. This strange dualist sect which became entangled with the southern Templars – especially at the end, when both Cathars and Templars were massacred at the siege of Montségur in 1244 – must have reached France from overland crusading journeys passing through the Balkans and the homeland of the Bogomils in Bosnia, whose religion closely resembled that of the Cathars.

Thus, in spite of 300 years of potential contact, there was little noticeable impact of Islamic Spain on southern France. Powerful aristocrats from the south such as Raymond IV, Count of

Toulouse, his son Bertrand and William IX, the Duke of Aquitaine (described as the 'first troubadour') all spent time in the Middle East, not in Muslim Spain. William IX spent a year at the court of Tancred in Antioch where 'he had without doubt the time to get to know the poetry and the customs of the Arabs' [Nelli, 1963, p41]. What changed at the end of the 11th century was that the Burgundian French came into contact with Islam as a direct consequence of the Crusades, starting with the First Crusade in 1095.

Knights Templar

It was the northern French, mainly from Burgundy, who were instigators of the Knights Templar and who became a means of transmission of Islamic influences into France and Europe. Southwestern areas like Roussillon and the Languedoc eventually had many Templars living in their regions – indeed, Raymond IV, Count of Toulouse, was a most influential member – but knights from the south were not part of the original Templar order of nine, most of whom were based in Champagne and Burgundy. The siting of the first Gothic cathedrals in northeast France, where many Templars came from, also suggests that the influence of Islam came via them from the Middle East, not from the south via the Languedoc.

In 1104, five years after the fall of Jerusalem to the Crusaders, Hugues Count of Champagne and his vassal Hugues de Payens made pilgrimage to the Holy Land together [Hancock, 1993, p95]. One can only speculate whether their motives were strictly Christian or, as seems more likely, they had received interesting reports after the First Crusade. Hugues de Payens later returned to Jerusalem alone and, in 1118, with permission from Baldwin II the Crusader king of Jerusalem, he and eight other knights set up the Templar Order on the Temple Mount, with himself as Grand Master. Hugues, Count of Champagne, did not himself join the order until 1125.

One man who never actually visited the Middle East but who provided the link between the Templars and the Gothic cathedrals was St Bernard of Clairvaux. It was from Hugues, Count of Champagne, that Bernard had received the land for his Cistercian monastery at Clairvaux. His uncle, André de Montbard, was also one of the original nine knights. And it was Bernard who sponsored the ordinances for this unique order of monastic soldiers in 1119 (confirmed at the Council of Troyes in 1128) which gave them exceptional autonomy, being exempt from the king's taxes and answerable only to the Pope. One of St Bernard's protégés, Pope Innocent II, made this clear in 1139 [Picknett & Prince, 1988, p126]. Without St Bernard's protection it is unlikely that the Templars would have been as successful as they were in pursuing controversial ideas relating to ancient civilization.

Undoubtedly, Bernard was a driving force behind the Gothic style. By 1130 he had persuaded his friend Henri Sanglier, the Archbishop of Sens, to rebuild his cathedral in the new style – that is, the Gothic style. This cathedral was the first entirely Gothic building in France from its ground plan up [Strachan, 2003, p31]. St Bernard was close to Abbott Suger, who in 1144 officially inaugurated the new style with the rebuilding of the eastern apse of the Abbey St Denis outside Paris. He was also close to Geoffrey, Bishop of Chartres, who rebuilt the western portal at Chartres in the new style in 1145 – the rest of the cathedral was restyled in Gothic only from 1194 after a fire [Strachan, 2003, p20].

St Bernard died in 1153 before the rest of Chartres was transformed; so its final appearance did not necessarily reflect his views since he believed that churches should be without embellishment. It is therefore difficult to know to what extent he connived in discoveries of the Templars that challenged the Church. He was, after all, one of the most highly respected churchmen of his day, going to the Languedoc at the request of the Pope in 1145 to dissuade the Cathars from their heresy and although he reported back that, were the Pope to 'examine their mode of life, you will find nothing irreproachable' [Strachan, 2003, p116], he still believed the Cathars to be doctrinally in error. Even so, did he take an active interest in the Middle East – not just in Jerusalem – because he wanted to acquire ancient knowledge and actively encourage the Templars in this pursuit?

Edessa, Harran and the Sabians

One reason for wondering about his motives was that, significantly, it was St Bernard, not the Pope, who preached the Second Crusade at Vezelay following the fall of Edessa to the Seljuk Turks in 1145 – the first loss of territory since the First Crusade. At the time, Jerusalem was not under threat and, as writer Adrian Gilbert points out, Edessa was 'of little strategic value, and its defense stretched resources'. Edessa was on the wrong side of the Euphrates, in an enclave surrounded by Arab states [Gilbert, 1996, p191]. So, why try to regain Edessa after only 50 years of Crusader control?

Before the Crusades, Edessa had been the third largest Christian community in the Byzantine Empire, after Constantinople and Antioch. It had been the home of the 'Mandylion', a legendarily miraculous cloth with an image of Jesus (in a similar tradition to St Veronica and the 'true icon'). Edessa's further claim to fame was that it could have been the famous Ur of the Chaldees, the birthplace of Abraham – known today as Sanliurfa. But was the real attraction of Edessa its close proximity to Harran, the northern Syrian safe haven for those with knowledge relating to ancient civilization?

Harran was unusual in the 12th century in being a city that was not Islamic, Jewish or Christian. Its main temple, eventually destroyed by the Mongols in 1259, was dedicated to the Mesopotamian Moon god, Sin. The city was a station on the old Silk Road. It is mentioned in the Bible for trading with the Phoenician city of Tyre in 'choice garments, in clothes of blue and embroidered work, and in carpets of colored stuff, bound with cord and made secure', along with places like the Garden of Eden and Assyria [*Ezekiel* 27:23-24]. It was also well known for its metallurgical trade, linked with mines in Asia Minor and Persia. This connection with metal, however, had another side to it.

Harran was most famous as a centre of alchemy, as practiced by Sabians who regarded Hermes as the founder of their school. The Harranian school was distinctive because of its 'sole use of minerals and metals in the operations', especially copper, which in writer Jack Lindsay's view indicates an ancient tradition, possibly dating to 1200 BC or before, when copper was the chief metal. The Sabians' beliefs were very old, with links to Egypt. The Victorian writer Ignatius Donnelly stated that 'In the Qarmus (the Koran) it is said of the Sabeans that they were of the religion of Noah' [Donnelly, 1950, p175].

Although the Arabs seized the city between AD 633 and 643 [Gilbert, 1996, p68], the Sabians carried on their cult while those around them gradually converted to Islam. Even so, the precariousness of their existence is illustrated by the account of the Caliph of Baghdad's visit to Harran in AD 830. He passed through Harran while fighting campaigns to the west and wanted to know if those who dressed differently from Muslims and Christians were 'people of the book' (that is, Muslims, Jews or Christians). Fortunately for them, he accepted their reply which was that their 'book' was the *Hermetica*, their prophet was Hermes and they were the Sabians referred to in the Koran.

Little did he realize how much they had contributed to his own culture, Sabians having moved to Baghdad when it was founded in AD 762 and having helped turn that city into a seat of learning and culture.[45] Indeed, they were a 'conduit for the transmission of ancient wisdom to the Arabs' [Gilbert, 1996, p70], especially to sects like the Sufis and the Druze. The academy that the Sabians established at Harran lasted until the 10th century, attracting initiates from as far afield as India (Sind), Damascus,

Tyre, Hieropolis, Heliopolis and Balkh [Lindsay, 1970, p314].

Contact with Templars

The Templars knew the Sufis and the Druze, and an especially fierce group of Ismaili fighters known as the *Hashishim* (or Assassins), employing them as secretaries, spies and translators while based in Jerusalem and on campaign [Baigent & Leigh, 1997, p82]. Their contact with the Sabians is revealed through an unexpected religious view that the Knights Templar held: the veneration of St John the Baptist in preference to Jesus.

The Templars were known to celebrate the feast day of John the Baptist on 24th June, adopting his symbol of the Lamb of God as their own. A Templar manuscript called the *Levitikon*, which explains the rituals for the nine grades of the Templar Order, contains a medieval version of *St John's Gospel* that describes the Templars as being of the 'Church of John' and in which they refer to themselves as 'Johnites' or 'original Christians'. In this manuscript Jesus, referred to as 'Yeshu the Anointed', is presented as an initiate of the mysteries of Osiris, and it claims that Paul and others founded the Christian Church without knowledge of Jesus' secret teachings [Picknett & Prince, 1988, pp189-91]. Their preference for John over Jesus was, however, unusual and potentially heretical.

What is plausible is that the source for this preference were the Sabians. As late as the 17th century Jesuit missionaries who encountered Sabians in southern Iraq referred to them as 'Christians of St John' [Schonfield, 1984, p131]. The Sabians acknowledged Jesus, whom they called 'Yeshu the Messiah', as one of them, but who had betrayed their secrets. They claimed that John the Baptist was their greatest leader, greater than Jesus because 'he also partly reflects aspects of the Babylonian god Oannes' [Schonfield, 1984, p132]. They also had descriptions of a similar Nativity to that in the Bible, but with John rather than Jesus as the holy infant.[46]

Knowledge from the East

If the Templars developed their ideas about Christianity from the Sabians, did they also learn from them about alchemy and other matters? There is little doubt that the Templars changed their entire approach to religion after being in the Middle East. They moved away from focusing on punishment and suffering: their own cross was the *croix pattée*, the more ancient equal-armed cross, and their religious buildings were circular, not cruciform.

Chartres may not have been built by the Templars themselves, but analysis of the cathedral reveals their influence: their cross, for example, the *croix pattée*, can be seen in stone carvings in the cathedral. They could have been the reason for the reference to forbidden astrology in the zodiacal stained-glass window at Chartres – unexpected in such a prominent ecclesiastical building.

There are also hints of alchemy in the statue of Melchizedek, who is holding a chalice, possibly containing the Philosopher's Stone, and in the blue of the stained-glass windows – only achievable through an alchemical process. Most intriguing is the alleged existence of a lodestone once suspended over the

labyrinth in the middle of the cathedral. The labyrinth is still there but both the copper plate and the lodestone have gone. This basic device, however, would have created an electrical field when set in motion – reminiscent of the field created for the ancient pharaohs [Gardner, 2004, p287].

The Chartres labyrinth

The Alchemical Grail

We can only really find out what the Templars learnt from the Sabians and others through circumstantial evidence, as they were highly secretive; their motto being 'to will, to dare, to know and to keep silent' [Sadhu, 1962, p471]. Some insight can be gained from coded references to alchemy in the Medieval Grail legends that began to circulate Europe in the 12th century. The first of these, Christian de Troyes' *Lis Contes du Graal*, first appeared in 1178, 60 years after the Templar order was created. It is no coincidence that its author came from Troyes, the Burgundian city close to where Hugues de Payens, first grand master of the order, had his lands. The later 13th century *Parzival* of Wolfram von Eschenbach even explicitly mentions the Templars in his version.

While these tales owe something to the Arthurian romances of knightly quests, they have only superficial Christian overtones. St Bernard's Cistercian monks were responsible for another version, *Queste del San Graal* (1190-1230) with a stronger Christian bias [Godwin, 1994, p82], and yet they are all clearly inspired by Templar experiences in the Middle East. The legends incorporate an undeniable Egyptian influence, often using the language of alchemy. Perceval, for instance, is referred to as the 'son of the widow', an epithet applied to Horus as the son of Isis, the widow of Osiris [Laidler, 1998, p182, p199]. The main theme of restoring to health a wounded ruler so that his land might also be healed has strong parallels with the Osirian rites and the *Heb Sed* ritual of the pharaohs. In the tale of *Gawain and the Green Knight*, the quest of Gawain specifically entails restoring the land to fruitfulness [Baigent & Leigh, 1997, p55].

Of particular interest are the Grail references to the Philosopher's Stone and the phoenix, the Egyptian *bennu* bird. In von Eschenbach's story, Parzival meets a hermit who tells him of Templar knights living on the wild mountain, the *Munsalvaesche*, who 'live from a stone whose essence is most pure', and which had special powers of eternal youth. This stone, according to the hermit, is called the *lapsit exillis* or 'the Gral'. The hermit explains that'such powers does the stone confer on mortal men that their flesh and bones are soon made young again'. He also says that 'By virtue of this stone the phoenix is burned to ashes in which he is reborn. Thus does the phoenix molt its feathers! Which done, it shines dazzling bright and lovely as before!' [von Eschenbach, 1980, p239].

It is understandable that the Templars would hide references to alchemy in the form of allegorical tales, as its practice remained a capital offence even in the lifetime of Sir Isaac Newton, one of its greatest practitioners in the 17th century.[47] It was not just their interest in alchemy that they kept to themselves: the rules of architecture that the Templars learned from the East were passed on in secret through guilds of stone-masons. Villard de Honnecourt was perhaps the first to write them down around 1235, 100 years after the new style first appeared in Europe [Heyman, 1995, p139].

Even more peculiar was the secrecy of the Templars about an essential part of their creed – the importance they attached to the feminine principle, which they referred to as *Sophia* or Wisdom. This desire for secrecy gave rise to strange rumors that the Templars worshipped a disembodied head called *Baphomet*. But when *Baphomet* is translated using the Essenes' Atbash cipher, it turns out to be the name Sophia [Schonfield, 1984, p164].[48] It is possible that they disguised their concept of feminine wisdom because, in the view of the Church, it risked association with the heretical Gnostics, one of whose texts buried at Nag Hammadi was entitled the *Pistis Sophia* [Pagels, 1979, pp84-5].

Nevertheless, their caution in this regard is odd. From the end of the 11th century there was a widespread interest, promoted by troubadours in southwestern France, in another Arabic influence – the ideal of courtly love [Nelli, 1963]. In addition, the Templars' sponsor, St Bernard, was candid in his support for the feminine principle. He not only encouraged a cult of the Madonna, preaching many sermons on the subject but, even more radically, he promoted Mary Magdalene, referring to her as 'the forgiven one'.

The Renaissance Challenge

This medieval secrecy of the 11th-12th centuries was in distinct contrast to the openness of the Renaissance in the middle of the 15th century. Times were still dangerous, but the challenge to the Church was more effective with powerful patrons such as the elder Cosimo de' Medici, whose family dynasties controlled the papacy anyway. A trend toward secularization had begun at the end of the 13th century and continued in the 14th century with figures like Dante, Petrarch and Bocaccio writing in the vernacular rather than in Latin [Baigent & Leigh, 1997, p109], opening the way for knowledge to become publicly available beyond the Church. Furthermore, the catastrophic effect of the Black Death during the 14th century had shaken faith in the Church, given that no amount of religious fervor had appeared to appease the wrath of God.

Another difference with the 12th century was that the last bastion of Christianity in Asia Minor disappeared in 1453 with the fall of Constantinople to the Ottoman Turks. There had been a rapprochement in 1438 between the Orthodox and Roman Churches (they had excommunicated each other in 1054 – a situation that prevailed right up until 1963) because of the need for protection from the Turks. The Byzantine emperor and the patriarch of Constantinople had come to Italy for help, and a council of scholars and ecclesiastics was held in Florence. When Constantinople fell, many scholars returned from there to Italy as a result of previous contacts.

These contacts were useful to those like Cosimo de' Medici who had dispatched missions to the Middle East to find ancient knowledge beyond the Church's teachings, such as the philosophy of Plato. It was not until 1460, however, that Cosimo got what he wanted – the *Hermetica* – which had survived in Asia Minor. By then Cosimo was old and knew he did not have long to live. He had already commissioned Marsilio Ficino to translate Plato, but he had Ficino turn to the *Hermetica* instead, establishing the Villa Careggi as an academy outside Florence where Ficino, Pico della Mirandola and others could dedicate themselves to the study of hermetic wisdom.

Cosimo died in 1464, one year after Ficino finished the *Hermetica* [Baigent & Leigh, 1997, p110]. The *Hermetica* was published more widely in 1471 and Ficino completed his translation of Plato in 1484 [Merkel & Debus 1988, p80]. Luckily, Cosimo's academy outlived him, helping to inspire so many aspects of the Renaissance with its restatement of the principles of ancient civilization from the *Hermetica*.

Hermeticism was a powerful challenge to the Church, not least because it meant placing the Sun at the centre of life, rather than God. According to Giordano Bruno, even Copernicus, a contemporary, attributed his heliocentric theory to hermeticism – Bruno said Copernicus quoted 'Hermes Trismegistus on the Sun as a visible god (from the Hermetic *Aesclepius*)' [Merkel & Debus, 1988, p21].

Thus a wide variety of talismans began to appear in all forms of art – from painting and buildings to gardens – trying to attract the power of the Sun, using either gold or plants sacred to the Sun and to draw down celestial influences ('as above, so below') [Baigent & Leigh, 1997, p114]. The Hermetic influence was especially strong in the revival of architecture and urban planning. Tommaso Campanella's concept of the City of the Sun (*Citta del Sole* or *Civitas Solis*), which was never built, was the most extreme example of this change.

Architects like Leon Battista Alberti (1404-72) or Palladio (1508-80) were well aware of the laws of cosmic order and of 'mathematical ratios that determine the harmony in macrocosm and microcosm'. Writing in 1450, Alberti followed his Gothic predecessors in stating that it was 'not actual numbers but their ratios that are of importance' [Baigent & Leigh, 1997, p210]. He claimed that it was Plato's *Timaeus* that was his inspiration for using a circle in the building of churches. It was not just ecclesiastical constructions, such as Brunelleschi's famous dome in Florence, that were affected – buildings such as Shakespeare's Globe Theatre in London were also deliberately circular.

Theatre and sacred music dramas, from which opera developed, were considered a useful means for conveying information about hermeticism. Spencer's *The Faerie Queen* is one such play containing themes based on occult philosophy [Baigent & Leigh, 1997, p224]. The works of Shakespeare are further examples, with *The Tempest* as his most alchemical play (it even identifies Prospero as a *magus*). Many of his plays revolve around the theme of love in all its forms. In this respect he echoed Ficino, who wrote in his *Commentary on (Plato's) Symposium*, 'An act of magic is the attraction of one thing by another in accordance with a certain natural kinship… Common attraction is born of love. This is true magic… Acts of magic, therefore, are acts of nature' [Merkel & Debus, 1988, p87].

Brunelleschi's dome in Florence

The *Hermetica* Discredited

Ficino and della Mirandola were eventually stopped in their enthusiasm for hermeticism by the Church when they were 'forced to write the *Apologiae* of their magic theses' between 1486-87 [Merkel & Debus, 1988, p128]. Individuals like Galileo and Giordano Bruno were less fortunate, falling foul of the Catholic Inquisition – Bruno being one of the most famous hermeticists who died at the stake for his beliefs in 1600. Others such as Paracelsus, Kepler and John Dee (Elizabeth I's personal astrologer) continued their involvement with hermeticism by trying to ensure that their use of magic could not be denounced as witchcraft, in spite of practicing at a time of some of the fiercest witch hunts [Merkel & Debus, 1988, p129, p163].

Finally, more than a century after its publication, the Church was able to destroy the popularity of the *Hermetica* with the help of an English king. James I, a devout Catholic and keen to discredit the *Hermetica*, actively encouraged Isaac Casuabon in his attempts to do so. For Casuabon and the Church it was critical to end the interest in hermeticism. This magical world of correspondences between the macrocosm and microcosm, between divine and mortal, God and man, which included forbidden subjects such as astrology and reincarnation, entirely bypassed the Church.

The method that Casuabon chose in 1614 to undermine the *Hermetica* was skilful. He didn't refute its contents. Instead, he more subtly challenged its authenticity as a truly ancient document. What Cosimo de' Medici and others had wanted to believe was that their manuscripts were copies of such ancient wisdom, going back to the time of Noah, and that they rivaled the Bible. Somehow, Casuabon was able to show that the *Hermetica* was no older than the 1st century AD, originating in Alexandria [Cornelius, 1994, p3; Fowden, 1993, pxxii]. As a consequence, the popular appeal of Hermeticism declined and became restricted to those interested in esotericism.

But what Casuabon was unable to realize in the context of his time was that the *Hermetica*, with its central doctrine of cosmic sympathy, contained fundamental truths about the physical world. With the benefit of hindsight we can see that Casuabon was wrong. What he could not know was that ancient hermeticism was the same as the modern 20th century concept of quantum physics, albeit expressed in very different language. Underlying both is the same idea of *interconnectedness* – what the German poet Goethe at the beginning of the 19th century called 'elective affinities' [Baigent & Leigh, 1997, p252], or what the Egyptians referred to as *magia* (high magic). Only now are we beginning to grasp the implications of this idea of interconnection, which was such an important part of the original model of ancient civilization.

Chapter 19

Magic and Quantum Physics

It is plausible to suggest that the ancients had some idea about what we call quantum physics. Herbert Frohlich of Liverpool University described quantum mechanics as the study of high levels of energy wherein molecules vibrate in unison and reach a state of vibrational coherence. Through our present-day understanding of quantum mechanics, we now take for granted lasers, superconductivity and ultrasound. It could be said that the practical application of quantum physics is knowing how to organize the chaos of random particles so that they achieve Frohlich's coherence, as in a laser beam where all the photons behave exactly the same.

Knowledge of quantum physics is the only sensible, even if difficult to argue, explanation for the ancients' ability to build with megaliths and create giant pyramids. They must have known how to interfere with gravity at a sub-molecular level, rendering less solid stones weighing hundreds of tons, without having our theoretical framework to explain such things. Otherwise, we are either in the realms of 'little green men' from outer space helping us poor earthlings, or we are left with mysteries which can be accounted for only by the use of impossibly large numbers of slaves and improbable engineering techniques.

The usual explanation for the Great Pyramid – that the Egyptians made ramps to carry up all the blocks – is impractical. The ramps would have to have been impossible lengths, running to more than a mile if they were to have a useful gradient, and they would need to be adjustable and extendable to maintain a perfect slope, as the pyramid is built up. It is logistically unlikely that hundreds or thousands of slaves could pull large numbers of 50-ton blocks, or heavier, over a large distance, let alone up a slope with nothing more than rope made of hemp. It is pitiful to watch teams of modern volunteers develop hernias in experimental reconstructions of the building of Stonehenge or other prehistoric monuments in vain attempts to prove an unviable explanations.

As writer Graham Hancock points out, even today with all our advanced technology there are few modern cranes capable of picking up more than 20 tons [Bauval & Hancock, 1997, p28]. In spite of modern lifting equipment and our technological advance, we still use girders for strength and poured concrete for bulk instead of the stone they used.

Excavations undertaken in the area around the Great Pyramid in the late 1980s and early 1990s revealed that those who worked on the pyramid were not necessarily poorly treated slaves. Quite the contrary, pyramid workers were well housed and fed. There is evidence of entire communities who lived with their families and worked in the area for parts of the year. Archaeologists have found evidence of 'bread production on a truly vast scale', as well as beer production. What is more surprising is the discovery of 'the remains of many choice cuts of prime meat', an expensive luxury in ancient Egypt.

Furthermore, analysis of skeletons found in the workers' graves shows that they received successful medical treatment for fractures and, in one case, even for amputation – all of which had healed before death [Tyldesley, 2000, pp54-59]. Comparison of the skeletons of the workers with those of nobles shows that, while the pyramid builders might not have lived as long as the nobles, they were nevertheless well looked after – not what you would expect for slaves.

Skills at a Smaller Scale
Not only were the ancients able to move around blocks of stone

that we would find difficult or impossible today, but also they expertly worked stone to produce the results they wanted. In both South America and Egypt there are examples of blocks cut deliberately to incorporate a corner, or shaped to fit older weathered stone and so to re-face it. This was the case with the Valley and Sphinx Temples, with granite ashlar restorations that perfectly covered the older, eroded limestone underneath [Schoch, 1999, p36]. They were not daunted by types of stone, either. Indeed, the harder the stone – granite, basalt, diorite – the more they were likely to use it.

So easy was it for them to work in stone that they could go to enormous trouble to create ornamental detail to make it look like wood or other living materials. The Step Pyramid complex at Saqqara in Egypt has many such details. The roof of the entrance corridor to the whole complex was carved to resemble palm logs and a fake door was made to look as if it is made of cedar wood [Rohl, 1998, p363]. In its inner court for the *Heb Sed* festival there is a row of shrines on the western side which have been deliber-ately built in stone to imitate *mudhif* reed buildings with curved roofs, that are still found today among the Marsh Arabs of southern Iraq (where the Mandaeans had once lived) [Rohl, 1998, p373].

These skills do not apply just to large monumental works but also to smaller items of more personal or ritual use, using the same hard stones. There are beads for personal jewelry composed of these materials with holes finer than modern needles, as well as jars, statues and other items. Some of the stone sculpture is of a quality and accuracy that would be difficult to produce today with modern techniques. One can see many examples in the British Museum of perfectly-shaped alabaster bowls and stone jugs with narrow necks for which ultrasound drilling production techniques are a plausible explanation.

Even so, how could something as 'scientific' as quantum physics have any connection with a notion as 'unscientific' as magic? First, there is an issue of semantics. 'Magic', as pointed out earlier, does not mean just the tricks of the illusionist, but the principle of attraction, as in the drawing down of a cosmic force by using the theurgic arts to interact with the gods by using 'suitable receiving instruments' [Lindsay, 1970, p52]. 'Magic' in this sense is hermeticism: it concerns the powers of attraction. Secondly, quantum physics does not exactly fit the usual definition of 'scientific' – it is an entirely new phenomenon in modern science even if it has been understood in the past in different terms.

The voyage of discovery into the world of sub-molecular particles has almost entirely undermined the classical physicist's position and attachment to an 'objective' scientific method. According to the Copenhagen Interpretation, as presented to the world by Niels Bohr in 1927, quantum physics is based on 'uncertainty, complementarity, probability and the disturbance of the system being observed by the observer' [Gribbin, 1991, p121]. Frohlich also stated that one of the qualities of uncertainty was the phenomenon of 'non-locality' – in other words, a particle could appear to be in two places at once [McTaggart, 2003, p63], and it is never clear where exactly it is.

Einstein was one of those who had difficulty in accepting the new physics. He said it was 'as if the ground had been pulled out from under one, with no firm foundation to be seen anywhere, upon which one could have built' [quoted in Capra, 1982, pp61-62]. Yet it was Einstein who, having published a paper on light quanta in 1905, was among the first to realize that an electron, or light in the form of a photon, could be both a particle and a wave.

Einstein was unhappy with this concept, which because he did not think it 'reconcilable with the experimentally verified consequences of the wave theory' [Gribbin, 1991, p81]. Niels Bohr, however, compared his idea of complementarity to the oriental symbol of *yin* and *yang* which he went on to adopt as part of his personal emblem. The problem for physicists was that they could never be sure which aspect of the photon or electron applied – wave or particle – thus giving rise to Frohlich's non-locality and Heisenberg's Uncertainty Principle.

According to the Uncertainty Principle, physicists could measure either the precise location of a particle or they could measure its momentum, but they could not measure both. Thus, experiments to detect waves always found waves, and particle experiments always found particles. Worse, physicists could only rely on probability to work out what an electron *might* be doing – behaving either as a particle or a wave – when it was not being observed. The very act of making the observation had a direct impact on the electron's behavior. 'By choosing to measure position precisely, we force a particle to develop more uncertainty in its momentum, and vice versa' [Gribbin, 1991, p120, p160]. In quantum physics it is therefore no longer possible to maintain objectivity and separate the role of the observer from the process of observing. As Werner Heisenberg put it; 'What we observe is not nature itself, but nature exposed to our method of questioning' [Capra, 1982, p152].

The Nature of Nature

As physicists reconsidered their methods of questioning, discoveries about quantum physics led physicists to question the nature of nature itself. This has led to a profound reassessment of the meaning of material substances that we have taken for granted. In terms of the classical physics of Newton, there was an assumption that the universe consisted of solid objects surrounded by empty space, upon which some force such as gravity may or may not have exerted pressure (e.g. Newton's formula: force = mass × acceleration, or F=ma). What quantum physics has done, however, has been to go beyond even Einstein's famous equation of $E=mc^2$.

With the insight that quantum physics has given them, physicists have realized that $E=mc^2$ does not mean that energy converts into mass, but that mass and energy are essentially the same. Hal Puthoff and colleagues take the view that there is 'no such thing as mass: only electric charge and energy, which together create the illusion of mass'. $E=mc^2$ is better understood as 'a statement about how much energy is required to give the appearance of a certain amount of mass'.[49]

Einstein knew from his work on relativity that matter couldn't be separated from its field of gravity and that gravity cannot be separated from curved space – matter and space are therefore inseparable. Puthoff also said, 'We may regard matter as being constituted by the regions of space in which the field is extremely intense. There is no place in this new kind of physics both for the field and matter, for the field is the only reality' [Capra, 1982 p231, p233].

What later physicists like Hal Puthoff are saying is that there is no such thing as empty space – even a vacuum is not devoid of sub-molecular activity – because there is an all-pervasive background electromagnetic radiation which they term the Zero Point Field (ZPF – zero, because even at absolute zero temperature there is still activity), which accounts for the phenomena of gravity and inertia [Haisch, Rueda, & Puthoff, 1994]. In other words, a solid object only gives the *appearance* of being solid. In fact, at the subatomic level it consists of electrical charges that operate at particular frequencies interacting with the background electromagnetic field, the ZPF, in terms of either repulsing or attracting, as either inertia or gravity.

Ernst Mach hinted at this interpretation with his principle, which stated that the inertia of a material object was not an intrinsic property of matter but a 'measure of its interaction with all the rest of the universe' [Capra, 1982, p231]. As physicist Fritjof Capra poetically described it, at a subatomic level, everything, including all mass, is a dynamic pattern of energy, 'probability patterns, interconnections in an inseparable cosmic web', an endless cosmic dance [Capra, 1982, p225]. These descriptions almost exactly parallel historian Brian Copenhaver's definition of Hermeticism. Quantum physics is arguably a re-statement in modern terms of the ancient hermeticism that only came to light again with the Renaissance.

The Renaissance – a Hermetic Revival

Copenhaver describes the Hermetic cosmos as 'an organic unity whose parts affect one another as participants in the same life'. He goes on to ascribe the unity of the cosmos to 'the physical agency of *pneuma* [breath] or *dynamis*, referred at other times to the ability of immaterial *nous* [intelligence] to penetrate all that exists' [Copenhaver, 1988, p81].

What they rediscovered in the Renaissance – thanks to the brave efforts of Cosimo de' Medici, Marsilio Ficino, Giordano Bruno and others – was something of what the Egyptians took for granted. They found out about the 'mystical force' of the universe, the *heka* that was the responsibility of the Egyptian priest-magician, the *Ur-Hekau*, to transmit to the other priests [Clark, 2000, pp358-9].

We might restrict our awareness of it to electromagnetic fields and particle behavior, but for the ancients and during the Renaissance, it was an all-pervasive concept, affecting every aspect of life. It was the essence of their relationship with the Divine. But while the hermeticists of the Renaissance may have understood the existence of a 'mystical force', what they didn't find out was how the ancients knew how to manipulate this force – how to use ancient quantum physics in any practical sense. That part had to wait until the 20th century. The Renaissance was therefore a time when interest in this mystical force remained at a more spiritual level, given full expression in diverse areas such as architecture, art, music and medicine, in an attempt to reconnect with the Divine outside the confines of the Catholic Church.

Botticelli's *Primavera* is a perfect illustration of this kind of thinking. *Primavera* is essentially a talismanic painting in which, by placing Venus at the centre, Botticelli endeavors to evoke the powers of that planet to inspire love. But in order for this to happen, first the god Zephyr as the *pneuma* breathes the Cosmic Spirit into Chloris, the world of winter, who then transforms into Flora, spring. At this point Cupid blindly fires his dart at the encircling three Graces – Beauty, Chastity and Passion – and from them, Hermes returns the Cosmic Spirit back to the Cosmos [Baigent & Leigh, 1997, p206].

Detail of the three graces from Botticelli's *Primavera*

Renaissance scholars like Marsilio Ficino believed that, through natural magic, through the 'invocation of Intelligences (celestial and demonic)', the 'hidden virtues or qualities in all things, plants, gems, animals' could be employed in (for example) medicine. Based on this belief Ficino wrote three books – medical, magical and astrological – on 'obtaining life from the heavens'. Music was also important for Ficino, and his Orphic Hymns were 'a system of invocatory planetary music' [Cornelius, 1994, pp3-4].

Indeed, a 'whole science of correspondences was developed: a 'natural magic' relating plants, herbs, stones, symbols and temperaments to the planets' [Gilbert, 1996, p73]. Cornelius Agrippa, writing in the early 1500s, later than Ficino but along the same lines, stated that 'The philosophers, especially the Arabians, say, that man's mind, when it is most intent upon any work, through its passion, and effects, is joined with the mind of the stars, and Intelligences, and being so joined is the cause that some wonderful virtue be infused into our works'. This was a magic of the Intelligences which followed Plato's view on synderesis, that it was the conscience, the highest part of the soul, that connected with the Divine [Cornelius, 1994, p310].

Like Attracts Like

It seems to me that Frohlich's modern concept of non-locality is present in the ancient idea that 'like attracts like', which was common knowledge among the ancients. There are several references in the Old Testament to this principle [*Jeremiah* 12:13, *Proverbs* 22:8 and *Job* 4:17, cited in Lindsay, 1970, p196]. Writer Jack Lindsay quotes Sextus as saying, 'The physical philosophers have a doctrine of high antiquity that like things are capable of knowing one another'. Plotinos was another Classical commentator who believed that magical practices are explained by sympathy. In general the view was that the 'main bias of the doctrine is towards Hermes, with his insistence that like begets like'. The ancients were also fond of quoting a text attributed to the 5th century BC Demokritas [Lindsay, 1970, pp6-7, p196].

In this text, Isis 'the Prophetess' tells her son Horus to go and find a 'certain laborer [called Achaab] and ask him what he has sown and what he has harvested, and you will learn from him that the man who sows wheat also harvests wheat, and the man who sows barley harvests also barley'. The text continues with the view that anything produced 'against the order of nature... [is] engendered in the state of a monster and cannot subsist. For a nature rejoices another nature, and a nature conquers another nature. Wheat engenders wheat, man engenders man, and similarly gold engenders gold' [Lindsay, 1970, p194]. Underlying this allegorical tale is the idea that the principle of attraction operates on many levels, although not necessarily in a deterministic sense.

In recognizing the process of interaction between the earthly and celestial spheres – that 'physical causality begins in the stars and is transmitted through the spheres to the Earth' [Copenhaver, 1988, p81] – Ficino and others still acknowledged a concept of free will. There was a place for individual decision-making. The *magus* had a role in the use of certain kinds of magic and talismans to draw down the benign influence of the cosmic forces. Nowhere was this acknowledgement of free will more apparent than in their attitude to astrology.

Renaissance interest in astrology thus rekindled an approach to the subject that predated the determinism of the 3rd century BC Greek Ptolemaic tradition. Ficino took the view that 'one could avoid the malignity of fate' and not be a passive victim of a birth chart [Baigent & Leigh, 1997, p115]. He had a strong belief in guardian angels or spirits and that 'every person has at birth one certain *daemon*, the guardian of his life, assigned by his own personal star, which helps him to that very task to which the celestials summoned him when he was born' [Cornelius, 1994, p191]. Interestingly, in his reference to *daemon* he was reviving its benign pre-Christian meaning from the Greek. Ficino was not alone in his views.

Even as early as the 12th century Guido Bonatus had begun to relearn from Arab sources about the more ancient method of casting horoscopes known as *horary* astrology. It was he who became a primary source for a later European revival of interest in it. Although it was said that the 'validity of particular instances of horary will depend on the character and quality of the prompting *daemon*' [Cornelius, 1994, p122], there was still the possibility of free will in this method. This interest in horary continued until the 17th century.

There is an illustrative story about horary regarding a famous 17th century master of the practice in England called William Lilly, who lived in Walton-on-Thames (he was known to Elias Ashmole, who edited his diaries) [Cornelius, 1994, p155]. In 1638 Lilly recorded details relating to a case of fish that he heard had just been stolen from the riverside near where he lived. He immediately cast a horoscope which showed a preponderance of water signs (not surprisingly) and that the thief had close associations with water. But, more significantly, the signs showed someone about to move house. Lilly made enquiries and a week later went with a constable and a barge-man, and together they found the culprit and the fish [Cornelius, 1994, p106].

Newton

By the 17th century, however, at the time of Sir Isaac Newton, the Church had reasserted its stranglehold, and the concept of cosmic sympathy had become part of occult lore again. Even in the 19th century the Church remained reluctant to give up its control of knowledge: witness its fuss about Darwin's theory of evolution and Champollion's translation of the Rosetta Stone; just in case the Bible's version of events was threatened.

In spite of life-threatening opposition from the Church, the classical physics of Sir Isaac Newton and others from the Renaissance onward was successful in correcting such fundamental nonsenses as ecclesiastical geocentrism or flat Earth beliefs. And yet, Newton was interested in a synthesis of knowledge which he believed had once been in the possession of humankind, that he referred to as *Prisca Sapienta* [White, 1997, p106], and had the courage to maintain a secret interest in illicit alchemy throughout his life. (It is perhaps ironic that Newton became Master of the Royal Mint when one reason for the ban on alchemy was to prevent people from illegally producing their own gold coinage!) Given Newton's obsessions and access to ancient manuscripts, it begs the question to what extent his work on gravity or light was entirely original and how much it might have owed to rediscovering information known by the ancients.

It is also difficult to know whether or not his work might have developed in a different direction had it not been for the oppression of the Church. It may be that Newton felt that it was easier to explain to the Church a mechanistic view of nature. Nevertheless, he did indicate privately that underlying processes might not necessarily be mechanical. He thought that there might be other principles at work, such as the idea of an 'attraction at a distance'. He would also have been aware that some alchemists 'believed that matter and spirit were interchangeable and imagined what they called a "Universal Spirit"' [White, 1997, p206].

By following the changes in Newton's thoughts on gravity between 1672 and 1687 it is possible to observe him gradually coming to 'perceive gravity as operating by action at a distance, made possible by a form of active principle' [White, 1997, p207]. In an unpublished conclusion to his masterpiece *Principia*, he makes clear his view that a theory of attraction could be applicable to the microcosm as much as to the macrocosm:

> For, from the forces of gravity, of magnetism and of electricity it is manifest that there are various kinds of natural forces... It is very well known that greater bodies act mutually upon each other by those forces, and I do not clearly see why lesser ones [i.e. those 'as yet unobserved'] should not act on one another by similar forces [White, 1997, p225].

Similarly, in his published work *Opticks,* he raises a number of potentially controversial opinions in a series of questions, including one which particularly anticipates quantum physics. In his Query No 5 he asks, 'Do not bodies and light act mutually upon one another; that is to say, bodies upon light in emitting, reflecting, refracting and inflecting it, and light upon bodies for heating them, and putting their parts into a vibrating motion wherein heat consists?" [White, 1997, p289] In expressing these views, Newton was following on from Hermetic ideas that had become so popular during the Renaissance, 200 years earlier.

Continuity of Knowledge
What happened in the 17th century, in effect, was that Casuabon and the Catholic Church delayed the rediscovery of the practical application of quantum physics by 300 years. Casuabon was not

even correct in disproving that the *Hermetica* was ancient. What he failed to realize was that, while the *Hermetica* manuscripts themselves may well have dated from the 1st century AD, the age of the manuscripts did not invalidate the possibility that the information contained within them was genuinely ancient. The Septuagint version of the Old Testament, for example, itself only dates from the 3rd century BC and was also constructed in Alexandria, and yet it relates to a much earlier biblical tradition.

Given the oral tradition of teaching that existed in the ancient world, it is not surprising that the *Corpus Hermetica* might not have had a written form before the 1st century AD. There is also evidence of sacred books attributed to Hermes Trismegistus being compiled especially for the benefit of the Greeks. One such early treatise, dating to the 2nd century BC, on astronomy, with zodiacal signs and references in the text to 'within, the concerns of Hermes...', was found in the temple library at Memphis [Baigent 1998, p195].

It is therefore plausible that the body of writings referred to as

the *Corpus Hermeticum* could have been collated in 1st century AD Alexandria from material based on much older knowledge. But Casuabon had no way of knowing about this older Egyptian knowledge because it was not accessible to Europeans until Champollion in the 19th century. The Egyptians were capable of consistently maintaining the same information continuously over thousands of years in different formats.

Champollion in ethnic dress

This consistency is evident in the comparison of a magical papyrus dating from the early 4th century AD, now in the Paris Bibliotheque Nationale (*P. Graec. Mag.IV*) with the Pyramid Texts of Saqqara, which are more than 4,000 years old. Both texts describe a similar rite, like that used in the Pharaoh's *Heb Sed* ritual, for obtaining a divine revelation by means of a spiritual initiation. The 4th century AD papyrus refers to the near-death experience of the initiate who has the same experience as the pharaoh of 'flying up' and becoming immortalized in a star, in order to receive communication with the Divine [Fowden, 1993, pp82-4] – just like the Pyramid Texts.

Apart from the enormous time gap, the major difference between the papyrus and the Pyramid Texts was that the former was written on papyrus in Greek and therefore was available, whereas the latter were in Egyptian hieroglyphs on the walls of the pyramids' inner chambers, remaining unexcavated until Gaston Maspero discovered them in 1881 [Naydler, 2005, p141].

It was only in the early 20th century that interest in the *Hermetica* began to revive – coincidentally about the same time that Einstein was writing about light quanta. In 1904 the first serious academic paper for a long time was published on the *Hermetica* by Richard Reitzenstein, entitled *The Hermetica and the Greek Magical Papyri*. Since then there has been debate about whether the *Hermetica* represented the hellenization of more ancient Egyptian views [Fowden, 1993, pp72-3] or a syncretic anthology fusing Egyptian and Greek thought, reflecting Egyptian influence on Hellenic thought (a view taken by Andre-Jean Festugière, writing in the 1940s) [Merkel & Debus, 1988, p45].

Progress

The Church has become increasingly marginalized with regard to science, reduced to opinions on morality rather than to any matters of substantial 'fact'. As a result, we now have very separate cultural categories for science and religion – though it is germane to my argument that this separation was not always the case. After all, the nature of belief is quite distinct from the experience of empirical observation.

These days the Church has claimed a spiritual and moral sphere of influence for itself, having backed off from its attempt to control the understanding of the physical world, which it has now left to science. However, currently there is occasional overlap in relation to bioethics – such as the treatment of embryos. What scientists now say to the Church is 'You get on with worrying about God. Leave us alone to concentrate on factual matters'.

Scientists have a point. Worrying about the existence of God – about what He thinks and what He wants – is, in many ways, a circular activity. It all rather depends on what religious people think God is supposed to do or, more particularly, what He wants us to do. What is more interesting is the question of how the nature of people's beliefs affect their attitude to free will and moral responsibility.

Science and religion still coincide in having a common philosophy of history based on a belief in 'progress'. The Church takes a teleological view of the past: we get progressively closer to God through moral action. Science regards every new laboratory discovery as another example of a step along the road toward a nebulous sci-fi concept of society.

While it might be true that beneficial changes do occur and genuine improvements are made, there are nevertheless problems with this kind of thinking. The constant reference to 'progress' every time there is a scientific breakthrough runs the risk of unrealistically raising people's expectations that there really is some perfect future world in which science has all the answers and it is just a question of funding research programs to unlock the keys of this perfect world. And yet, it can happen that some advances, for example in medicine, can turn out in the end to represent a step backwards. No doubt, there were many after WWI who were grateful for the discovery of penicillin but, equally, there are now thousands who die every year because the oversubscribing of antibiotics has allowed bacterial strains like MRSA to become antibiotic-resistant.

The Cartesian Split

There is an additional danger that 'progressing' toward an intangible like 'God' or 'perfect' science encourages the idea that the 'end' justifies the 'means'. While there is no doubt that the use of 'ends' to justify the 'means' can result in insensitivity, a lack of feeling is also essential to science's aim of objectivity. We have allowed the left side of our brains to dominate our thought processes because we value being logical and are scornful of any intuition or subjective emotion that cannot be externally verified. This is a peculiarly Western frame of mind which we describe as *Cartesian* (in the manner of Descartes). Scientists have been so keen on separating out objectivity that it has been thought that there are even separate neural systems for emotion and reason. The older brain core was presumed to be the governor of emotion, while the neocortex was traditionally thought to be the site of reason. Antonio Damasio, a professor of neurology, has, however, come to the conclusion that the neocortex 'becomes engaged *along with* the older brain core' and 'rationality results from their concerted activity' [Damasio, 1996, p251, p128].

What Prof Damasio's clinical experience has shown him is that it is actually more difficult for us to make rational decisions without emotion. He has treated many patients who, from whatever cause – through accident or disease – have had their brains physically damaged in such a way that they remain quite logical but have little ability to respond emotionally [Damasio, 1996, pp193-194, p51]. Sociopaths and psychopaths are, after all, examples of a 'pathological state in which a decline in rationality is accompanied by diminution or absence of feeling'. His work has suggested to him that 'feeling was an integral component of the machinery of reason', and that 'reason may not be as pure as most of us think it is... and that emotions and feeling may be enmeshed in its networks, for worse *and* for better' [Damasio, 1996, p178, pxiv].

The difficulty that we have created for ourselves in disregarding emotion in the ratiocination process is that we have misplaced our confidence in the objectivity of the scientific method. Because we discount emotion in our effort to be objective, we are then unable to assess what is referred to as the 'placebo effect' in experiments involving people. It makes no sense that the only scientific method for testing drugs that the medical profession will accept is the standard double-blind test – using sample groups with and without the drug, and a control group with a placebo.

This method has to assume laboratory conditions in order to be 'scientific' when testing medicines, so that statistically relevant comparisons can be made, even though no two people are alike. The need to minimize differences in the interests of achieving 'objective' results has distressing consequences for real people. It means that new pharmaceutical products have to factor in a statistically acceptable level of side-effects in order to take account of the inevitable dissimilarities between patients. Just as doctors lack sufficient time for patients, so the remedies they dispense to patients have no regard for them as individuals.

What is so hard for us is that this 'scientific' way of thinking has been ingrained in us for such a long time, long before Descartes. Descartes merely formalized an attitude which existed many centuries before him. The importance that we attach to reason and logic we can trace back at least to the ancient Greeks over 2,000 years ago. Meanwhile, our willingness to tolerate unpleasant side-effects in the pursuit of any goal we can, perhaps, attribute to the militaristic legacy of the Romans.

Hermeticism – the Future

So where do we go from here? How do we re-connect scientific knowledge and spirituality in a way that is relevant to us today?

Undoubtedly, shamanic experiences are not likely to be easily resurrected in modern practice – although we could stop treating the shamans that still exist in obscure parts of the world as mere anthropological curiosities and give their opinions greater respect. They might actually be able to find out useful information that could help us in our current predicament.

Other esoteric aspects of the model remain equally suspect to most people. Astrology is still far from respectable. We continue to hold the same negative attitude to astrology as Perkins in the 17th century: we merely substitute 'unscientific' for his 'work of the Devil'. As for alchemy, it is so little understood that it is practically irrelevant; although writer Michael Baigent points out that technical alchemy resulted in knowledge about distillation and heating to very high temperatures, with perfume and alcohol (another Arabic word, *al-kohl*) being the by-product of these processes [Baigent 1998, p211, p94]. In these two areas alone – astrology and alchemy – we remain strongly ambivalent about our ancestors' attitudes.

On the one hand, we revere Newton as the father of modern physics, and Galen, a 1st century AD physician from Alexandria, as the father of modern medicine. On the other hand, we dismiss certain beliefs of these men as being superstitious nonsense. Yet Galen was clearly an intelligent man who took the view that 'astrology was one of the greatest achievements of Egyptian astronomers', and that someone who 'refused to observe these things, or who refused to believe in the observation of others, was doubtlessly a sophist... who does not attempt to proceed from understood phenomena to hidden things' [Merkel & Debus, 1988, p35].

Perhaps it is about time we acknowledged that our ancestors were not backward pagan idiots and that they might have had a point – even if they got some things wrong. It is we, arguably, who have gone backwards: as the philosophers de Santillana and von Dechend say, 'Our ancestors of far-off times were endowed with minds wholly comparable to ours, and were capable of rational processes' [de Santillana & von Dechend, 1977, p68].

Even now, we have not fully understood the wider implications of quantum physics. The ambivalence regarding homeopathy is a case in point. When Prof Jacques Benveniste and others published their findings in *Nature* in 1988 regarding the 'memory of water' (homeopathy), there was a disgraceful and shameful attack on him from the scientific community. This included, bizarrely enough, the use of a 'professional' magician to re-run Benveniste's experiments to prove them wrong [McTaggart, 2003, p80]. Such extreme and desperate measures echo the earlier tactics employed by Casuabon in the 17th century – except that this time the orthodoxy that was threatened was not the Church but 'science'.

While it might be amusing to wonder how medical science will resolve the inconsistency between 'unscientific' particle physics and the mundane level of everyday 'objective' science, it is also sad that we have had to live with the limitations of mundane science for so long. If we had not been so successfully brainwashed and paralyzed with fear by the romanized Christians, maybe we might have retained this part of the original model of civilization, or we might have rediscovered it earlier.

The concept of the ZPF and the interconnectedness of particle behavior is beyond doubt the same as the ancient idea of cosmic sympathy. We now have the chance to reincorporate the paradox that quantum physics represents into our total understanding of what it means to be 'civilized'. We have an opportunity to re-examine all the other aspects of how we live from the same perspective.

Conclusion

What can we learn from the ancient past? We could stop thinking that 'civilization', in the sense of 'life in cities', is an idea that evolved naturally. I have tried to show that there was no inexorable trend from the end of the Ice Age onwards. Being 'civilized' is not an innate part of the human experience: it is an artificial construct that human beings have had to learn. We have been happy to co-exist in tightly packed urban spaces for thousands of years because we have come to value the benefits of the original archetype: such as the convenient docility of domestic animals who provide for us; or the advantage of a highly organised infrastructure that brings fresh water to our door.

But, however much we might still appreciate this unnatural way of living that we first experienced thousands of years ago, we fail to realise how much we have lost in moving so far away from the original model. We might call it 'progress' – because of our scientific and technological 'advances' - but we are in danger of deluding ourselves. Not only have we lost the secrets of the ancients' technology, but we have lost contact with their way of thinking. We no longer know how to live in 'truth', in *ma'at*, in harmony with our home, the planet earth.

Now, as we find ourselves on the threshold of an environmental crisis, arguably as severe as that faced by civilization at the end of the 4th Millennium BC, it could well be worthwhile reconsidering the original archetype. After all, the ancient civilizers were able to survive catastrophe and re-establish a city-based life without fundamentally changing its concepts. To a large extent it doesn't matter whether or not the climatic changes that we are beginning to experience are our fault or not. The point is that we are going to have find ways of dealing with their impact, especially on our basic needs for food, water and energy sources – and access to the ancients' zero-carbon technology could address one of those requirements.

We need to use science sensibly to work out how the pyramids were built and the megaliths were moved. Non-scientists have revealed enough clues that the technology of the ancients involved electromagnetism, ultrasound and sympathetic resonance – all without a carbon footprint or enormous structures such as the Large Hadron Collider outside Geneva. In a rather disconcerting parallel with medieval theologians arguing about angels on the end of a pin head, modern physicists seem to be more concerned about discussing obscure aspects of string theory than in solving practical problems. They need to stop wasting time and resources, and re-examine the ancient past with more respect.

Timing

Time is of the essence. What James Lovelock of Gaia fame and others have made clear is that the timescale is unknowable. Modern chaos theory explains how tiny changes in initial conditions can have exponential consequences which are entirely unpredictable. The tipping point could come sooner than we realise. One reason for the uncertainty lies in the difficulty in modeling the positive feedback mechanisms in the Earth's planetary system: for example, the melting permafrost in Siberia in itself releases even more methane and carbon dioxide into the atmosphere, thus amplifying any manmade contribution from the burning of fossil fuels. What is particularly worrying is that empirical evidence collected by earth scientists directly from the Arctic Ocean or wherever indicates that the rate of change is happening faster than the computer-based climate change models predict.

We already have unsettled seasonal weather with milder winters or wet summers, violent storms with flash floods and rising sea levels. In 2008, the government of the Seychelles, the popular tourist islands, started seeking land in another country in the knowledge that their own islands would sooner or later disappear.

We are told to expect more earthquakes, volcanoes and tsunamis. The effect of Hurricane Katerina on New Orleans and the 2004 tsunami were real 'wake-up calls' to the power of nature. In the worst case scenario we have been warned that Western Europe could lose the Gulf Stream. So much cold water could flow south from the Arctic as the polar ice caps melt that it would push the warmer Gulf Stream away, and Britain would end up with a climate like that of Newfoundland with long, hard winters. If Europe were to lose the Gulf Stream, our need for energy would then increase enormously because of colder winters.

Why worry?

So, what should we do? If making a personal decision to go to work by bike instead of the car makes little difference to the melting of the polar ice cap, or the release of swamp gas from Siberian permafrost, why bother? As a friend of mine once reasonably pointed out, if the world is going to come to an end, "why do the sensible thing and stock up on baked beans and mineral water? Much better to open the 'Petrus' and have a party". We might as well enjoy life while we can and reap the rewards of bigger, faster, easier way of living, driving high performance cars on pointless journeys and continuing to do all the other things that we know are environmentally destructive. If the inevitable is going to happen, why change one's lifestyle?

Judging by the way most people live their lives, this attitude of resignation seems to be commonplace even among intelligent and aware people. If there is little we can do to alter our fate, why worry, especially when any alternative is costly and inconvenient? And yet, there is one very good reason for radically altering one's way of living now which can be summed up in one word: *preparation*.

The difficulty with my friend's approach is that, appealing though it is, waking up with a hangover, nothing to eat, and only a vague memory of a fun time leaves one rather ill-equipped to cope, even on a short-term basis. A more useful response would be to prepare ourselves and our families mentally, emotionally and physically for what lies ahead, and take responsibility for our own actions, while at the same time letting our children develop their own sense of responsibility. The children who solve the problems of the future will be those who were allowed to climb trees *without* adult supervision. At present, the excessive concern with liability – both corporate and personal – and the resulting desire to become risk-averse, combined with the corrosive effects of welfare dependency, have allowed many in Western society to succumb to passivity.

An ability to assess risk for oneself might make all the difference between survival or not: we cannot rely on either God or the Health & Safety Executive to work it out for us. Indeed, one indirect effect of extreme chaos could be that the bureaucracy of 'Nanny' State would implode under the strain of attempting to micro-manage major crises. For those in the West with lifestyles based on cheap fuel and reliable electricity supplies – and who believe in the unrealistic marketing hype of a perfect consumerist future - life could become not just uncomfortable but frightening. From a practical point of view, we therefore need to think through the consequences of climate change for ourselves. A survival strategy in which enough citizens grew their own food and invested their own time and money in supplying some of their own energy, such as the micro-generation of eco-electricity through solar or wind, might be one way to avoid the worst excesses of a breakdown in society.

As my historical analysis attempts to demonstrate, 'progress' is not inevitable. It is possible for humanity sometimes to go backwards and for even previously civilized societies to resort to barbaric behaviour when under stress (viz the change from Bronze Age to Iron Age and the *molk* sacrifice). 'Progress' is as much a state of mind which can exist at any point in time. We have already begun to lose the civilizing arts of writing and cooking among the

318

Western urban poor, and a kind of tribal, feral culture is beginning to emerge in certain urban areas – an indicator, surely, that civilization is not genetically imprinted on humanity.

An Opportunity

In many ways chaos on the scale of the disaster of the 4th millennium BC would be a welcome opportunity for positive change, if it meant the chance to re-introduce civilization along the lines of the ancient model in all its aspects (with the addition, of course, of washing machines and fridges…). It would be bliss to be surrounded by buildings, art and music that subliminally incorporated classical proportions based on the 'Golden Section' (ϕ) and not to feel either shocked by a modern 'statement' or guilt for not understanding it. Above all, it would be amazing to live again 'in *ma'at*' and create urban spaces that acknowledged people's psychological and practical needs while at the same time reconnecting us with the natural world.

It could be said that modern cities already conform to a concept of interconnection as they consist of a myriad of interdependencies. They have become so successful in providing goods and services that people can specialise in different areas of work without worrying about who will satisfy their needs. The globalisation of trade is a further extension of this point.

But we have paid a price in our modern Western cities for a highly competitive international trading network that has effectively exported skills away from the West. Our cities have lost their function as places where manual trades - such as smithying or tanning or shoe making or spinning or weaving - can happen. Some how we need to give our cities back a sense of purpose based on meaningful work (rather than shelf-stacking or fast food service); and so help to address the powerful feeling of alienation and despair that exists in our urban centres and which is reflected in increasing levels of alcohol and drug abuse.

Indeed, Western cities have become such artificial microcosms, so separated from providing the basic means of life that people no longer know where their food comes from (eg that milk comes from a cow and not from a carton). Our cities and our businesses are dangerously disconnected from the natural environment; unlike the cities of the ancient world which were more conscious of their relationship with nature. Take water as one simple example. The ancients were well aware of the importance of recycling rainwater and built their cities accordingly; either incorporating vast catchment pavements for rain as in the Mayan city of Tikal, or a simple basin in the centre of a courtyard in the C6th BC Rome. We talk about the usefulness of such measures but we have yet to implement them.

Our experience of the natural world has become increasingly mediated through the centralised buying patterns of supermarkets which no longer reflect local produce or even seasons. When we do have contact with the natural world through that organised method of food production called 'farming' we still remain disconnected. We breed instincts out of animals in order to keep them in unnatural indoor units as mere protein providers that cannot cope with life in a natural environment - such as chickens grown so fast for their breast meat alone that their legs break under the weight of their bodies or dairy cows that have such enlarged udders they can barely walk.

In addition, the desire for efficiency through specialisation has taken animals out of the traditional rotation between crops and grazing. Livestock are then no longer available to supply natural fertiliser – thus increasing the demand for fossil fuel derived synthetic fertilisers. The drive to maximise yields has created such dependency on artificial means that the soil is already losing its natural fertility with the result that we are damaging the nutritional quality of our food.

A return to a non-industrial, 'civilised' approach to farming that respected the natural instincts of animals, as well as the earth's own ability to restore its fertility would have benefits for people as well as animals. If we ate more vegetables and less animal protein it would probably improve our health. We would suffer less cholesterol and lower cancer rates – not least because one cause of cancer could well be an overgrowth of cell building from too much protein.

Science & Religion
We could also take more care of own health by knowing how to treat ourselves for minor ailments, and so make less demand on health professionals. The system would then be able to respond better to a 'flu pandemic. Medical science is, nevertheless, bound to be restricted in its effectiveness while it clings to a 'scientific' method that is not reconciled with quantum physics, and which cannot treat people as individuals rather than as statistics. If medical science were not at least 100 years behind the rest of science in clinging to Cartesian logic, we could explore with more confidence alternative treatments, such as homeopathy, which are explicable through quantum physics. In the healing of people what matters is what works, not the method by which the treatment has been developed.

Key to the survival of the original archetype of civilisation over thousands of years was the strange relationship between science and religion, between the technical and the transcendental. Then, the priestly caste had all the answers and knew what to do. Above all, in their practical applications of *magia* – such as in the cutting and moving of giant stones – they did not lose their sense of reverence or their awareness that the power of nature could be manipulated for humanity's benefit but not totally dominated and controlled.

We now have the chance to create a new synthesis of religion and science, in the sense of belief and knowledge, that reincorporates the subjective and objective; with quantum physics as the ultimate in paradox (ie the *truth* that light is both a particle and a wave) pointing the way forward in terms of interconnectedness. A modern form of polytheism might also help us to see how the Divine manifests itself everywhere, thus echoing the understanding in quantum physics of the interconnectedness of the physical world.

For us monotheism is an experiment that has not worked – it has brought us misery, guilt, despair, and ignorance. Thinking back to the proto-Indo-Europeans, religion should be more about what you put in your *heart*, not your head; as the Egyptians indicated, it was the weighing of the 'heart soul' (the *Ab*) that mattered because that contained the record of the soul's integrity. A different understanding of the immortality of the soul might also help us deal with the terrible grief and separation of death and not see it as the end but as part of the cycle of life.

The Future

Whatever happens, we will carry on. It will not be the end of the world – just the end of the world as we know it. The challenge is to survive as 'civilised' people. Mankind has recovered from countless environmental disasters before with far less technology than we now have at our disposal. Provided that we can adjust our expectations and improve our practical skills, there is no reason why we should not continue. Now, while we have time, we can at least be aware of our vulnerabilities and our dependencies – and even if climate change had little effect in our lifetimes, or our children's, we could still try and live in a more civilised and less polluted world. We already know so much about what is important, what needs to happen (living in balance and harmony with the natural world and each other). Many of us are just waiting for that certain push to change our lives.

The purpose of this book is to contribute a better understanding of our historical and cultural background; of how we lost the knowledge of what it meant to 'live in *ma'at*'; and how civilisation almost recovered. Knowing more about our ancient past, and realising that it was about more than just exotic archaeology and strange funerary arrangements, could help us redress the imbalance of the Roman legacy and pernicious influence of religion. We can then face the future with a clearer mind. But we will only be truly civilised when we understand how to integrate the paradoxes (objective and subjective, male and female, particles and waves, etc, etc) and that chaos, in the sense of freewill or creativity, needs to be at the heart of order. Our fate is in our hands. We don't have to wait for circumstances to force change upon us. If we use our *nous* now we can avoid a fate worse than death, which would be to live inadvertently in a hell of our own making. We can start to live in harmony today, not wait for tomorrow. We have free will and we can choose to be civilised.

Constantly regard the universe as one living being, having one substance and one soul; and observe how all things have reference to one perception, the perception of this one living being; and how all things act with one movement; and how all things are the cooperating causes of all things which exist; observe too the continuous spinning of the thread and the contexture of the web.

[Emperor Marcus Aurelius, C2nd AD *The Meditations*]

Bibliography

Alford, Alan F, *Pyramid of Secrets*, Eridu Publishers, Walsall, 2003.

Allegro, John M, *The Dead Sea Scrolls and the Christian Myth*, Abacus, London, 1981.

Allen, J M, *Atlantis – The Andes Solution*, Windrush Press, UK, 1998.

Andrews, Richard & Schellenberger, Paul, *The Tomb of God – the Body of Jesus and the Solution to a 2,000 year old Mystery*, Little Brown, London, 1996.

Aubet, Maria Eugenia, *The Phoenicians and the West*, Cambridge Univ Press, 1993.

Bahn, Paul G & Vertut, Jean, *Journey through the Ice Age*, Weidenfeld & Nicolson, London, 1997.

Baigent, Michael, *From the Omens of Babylon: Astrology and Ancient Mesopotamia*, Arkana Penguin Books, 1994.

Baigent, Michael, *Ancient Traces – Mysteries in Ancient and Early History*, Viking, London, 1998.

Baigent, Michael & Leigh, Richard, *The Elixir and the Stone*, Viking, London, 1997.

Baigent, Michael & Leigh, Richard, *The Dead Sea Scrolls Deception – the Dead Sea Scrolls and how the Church conspired to suppress them*, Arrow Books, London, 2001.

Baillie, M G L, *A Slice through Time – Dendrochronology and Precision Dating*, Batsford, London, 1995.

Bakus, John, *The Acoustical Foundations of Music*, John Murray, London, 1969.

Baldick, Julian, *Animal and Shaman – Ancient Religions of Central Asia*, I B Tauris, London, New York, 2000.

Barber, Elizabeth Wayland, *The Mummies of Urumchi*, Pan, London, 1999.

Barclay, Olivia, *Horary Astrology Rediscovered – a study in Classical Astrology*, Whitford Press, Pennsylvania, 1990.

Bauval, Robert & Hancock, Graham, *Keeper of Genesis – A Quest for the Hidden Legacy of Mankind*, Arrow, London, 1997.

Berlitz, Charles, *The Bermuda Triangle – the Great Unsolved Mystery of Our Time*. Souvenir Press, London, 1975.

Berresford Ellis, Peter, *The Celtic Empire – the First Millennium of Celtic History*, Constable, London, 1990.

Berresford Ellis, Peter, *Celt and Roman – the Celts of Italy*, Constable, London, 1998.

Bradley, Richard, *The Significance of Monuments*, Routledge, 1998.

Brox, Norbert, *A History of the Early Church*, Continuum, Pennsylvania, 1994.

Buttery, Alan, *Armies and Enemies of Ancient Egypt and Assyria, 3200 BC to 612 BC*, War Games Research Group, 1974.

Capra, Fritjof, *The Tao of Physics*, Flamingo, HarperCollins, London 1982.

Castleden, Rodney, *Minoans – Life in Bronze Age Crete*, Routledge, London, 1993.

Chadwick, Henry, *The Penguin History of the Church: the Early Church*, revised edn, Penguin, London, 1993.

Chadwick, Nora K, *The Druids*, Univ of Wales Press, Cardiff, 1997.

Clark, Rosemary, *The Sacred Tradition in Ancient Egypt – the Esoteric Wisdom Revealed*, Llewellyn, Minnesota USA, 2000.

Collins, Andrew, *Gods of Eden*, Headline, London, 1998.

Copenhaver, Brian, *Hermetica*, Cambridge Univ Press, Cambridge, 1992.

Copenhaver, Brian, 'Hermes Trismegistus, Proclus, and a Philosophy of Magic', in *Hermeticism and the Renaissance*, ed Ingrid Merkel & Allen G Debus, Folger Books, Washington, London & Toronto, 1988.

Cornelius, Geoffrey, *The Moment of Astrology – Origins in Divination*, Arkana Penguin, 1994.

Cornell, T J, *The Beginnings of Rome – Italy and Rome from the Bronze Age to the Punic Wars*, Routledge, 1995.

Cowan, David & Silk, Anne, *Ancient Energies of the Earth – an extraordinary journey into the Earth's natural energy system*, Thorson, London, 1999.

Damasio, Antonio R, *Descartes' Error: Emotion, Reason and the Human Brain*, PaperMac, London, 1996.

de Santillana, Giorgio & von Dechend, Hertha, *Hamlet's Mill*, David R Godine Publ, New Hampshire, 1977.

Doane, Doris Chase, *Modern Horary Astrology*, Am Fed of Astrologers, 1994.

Donnelly, Ignatius, *Atlantis – The Antediluvian World*, rev ed, Egerton Sykes, Sidgwick & Jackson, London, 1950 (first edn 1882).

Dunbavin, Paul, *The Atlantis Researches: the Earth's Rotation in Mythology and Prehistory*, Third Millennium, Nottingham, 1995.

Dunn, Christopher, *The Giza Power Plant – Technologies of Ancient Egypt*, Bear & Co, Rochester, Vermont, 1998.

Dupont-Sommer, A, *The Essene Writings from Qumran*, tr G Vermes, Basil Blackwell, Oxford, 1961.

Fagan, Brian, *The Long Summer – How Climate changed Civilization*, Granta, London, 2004.

Feather, Robert, *The Copper Scroll Decoded – one Man's Search for the Fabulous Treasures of Ancient Egypt*, Thorsons, London, 1999.

Fitton, J Lesley, *Minoans, Peoples of the Past*, British Museum Press, London, 2002.

Fowden, Garth, *The Egyptian Hermes – a Historical Approach to the late Pagan Mind*, Princeton Univ Press, New Jersey, 1993.

Freke, Timothy & Gandy, Peter, *The Hermetica*, Piatkus, London, 1997.

Freke, Timothy & Gandy, Peter, *The Jesus Mysteries – was the original Jesus a Pagan God?* Thorsons, London, 2000.

Frend, W H C, *The Rise of Christianity*, Darton, Longman & Todd, London, 1984.

Gardner, Laurence, *Lost Secrets of the Sacred Ark – Amazing Revelations of the Incredible Power of Gold*, Element, HarperCollins, London, 2004.

Gilbert, Adrian G, *Magi – the Quest for a Secret Tradition*, Bloomsbury, London, 1996.

Godwin, Malcolm, *The Holy Grail – its origins, secrets and meaning revealed*, Bloomsbury, London, 1994.

Gorbunova, N G, *The Culture of Ancient Ferghana – VI century BC-VI century AD.* The Hermitage, Leningrad, tr AP Andryushkin, BAR Int'l Series 281, Oxford, 1986.

Gribbin, John, *In Search of Schrödinger's Cat – Quantum Physics and Reality*, Black Swan, Transworld, London, 1991.

Haisch, Bernhard, Rueda, Alfonso & Puthoff, H E, 'Beyond $E=mc^2$', in *The Sciences*, 1994.

Hancock, Graham, *The Sign and the Seal*, Arrow, Random House, London, 1993.

Hancock, Graham & Faiia, Santha, *Heaven's Mirror – Quest for the Lost Civilization*, Michael Joseph, London, 1998.

Harding, A F, *European Societies in the Bronze Age*, Cambridge Univ Press, 2000.

Harvey, Clare G & Cochrane, Amanda, *Principles of Vibrational Healing*, Thorsons, London, 1998.

Hatcher Childress, David, *Technology of the Gods – the incredible sciences of the ancients*, Adventures Unlimited, Illinois, USA, 2000.

Hawkes, Jacquetta, *The First Great Civilizations*, 1973.

Heyman, Jacques, *The Stone Skeleton – Structural Engineering of Masonry Architecture*, Univ of Cambridge, Cambridge, 1995.

Hoffmann, R Joseph, tr, *Celsus On the True Doctrine – A Discourse against the Christians.* Oxford Univ Press, New York, 1987.

Jones, Steve, *The Language of the Genes*, Flamingo HarperCollins, London, 2000.

Knight, Christopher & Lomas, Robert, *The Book of Hiram – Unlocking the Secrets of the Hiram Key*, Arrow Books, London,

2000.

Knight, Christopher & Lomas, Robert, *Uriel's Machine – the Ancient Origins of Science*, Arrow Books, London, 2000.

Laidler, Keith, *The Head of God – the lost Treasure of the Templars*, Weidenfeld and Nicolson, London, 1998.

Langone, John, *Superconductivity – the New Alchemy*, Contemporary, Chicago, 1989.

Leick, Gwendolyn, *Mesopotamia – the invention of the city*, Allen Lane, London, 2001.

Lindsay, Jack, *The Origins of Alchemy in Graeco-Roman Egypt*, Frederick Muller, London, 1970.

Mack, Burton L, *The Lost Gospel – the Book of Q and Christian Origins*, Element, Shaftesbury, 1993.

Maisels, Charles Keith, *Early Civilizations of the Old World – the formative histories of Egypt, the Levant, Mesopotamia, India and China*, Routledge, London, 1999.

Mallory, J P, *In Search of the Indo-Europeans*, Thames & Hudson, London, 1991.

Marazov, Ivan, ed, *Ancient Gold – the Wealth of the Thracians*, Harry N Abrams Inc, New York, 1997.

McCarty, Nick, *Troy – the Myth and Reality behind the Epic Legend*, Carlton, London, 2004.

McTaggart, Lynne, *The Field – the Quest for the Secret Force of the Universe*, Element, HarperCollins, London, 2003.

Merkel, Ingrid & Debus, Allen G, eds, *Hermeticism and the Renaissance*, Folger Books, Washington, London & Toronto, 1988.

Mithen, Steven, *After the Ice – a Global Human History 20,000 -5,000 BC*, Orion Books Ltd, London, 2004.

Mohen, Jean-Pierre & Elvere, Christiane, *The Bronze Age in Europe – Gods, Heroes and Treasure.* Thames & Hudson, London, 2000.

Morrison, Tony, *Pathways to the Gods - the Mystery of the Andes Lines*, Michael Russell, Salisbury, 1978.

Morton, Chris & Thomas, Ceri Louise, *The Mystery of the Crystal*

Skulls, Element, HarperCollins, London, 2003.

Narby, Jeremy, *The Cosmic Serpent, DNA and the Origins of Knowledge*, Phoenix, Orion, London, 1999.

Naydler, Jeremy, *Shamanic Wisdom in the Pyramid Texts – the Mystical Tradition of Ancient Egypt*, Inner Traditions, Rochester, Vermont, 2005.

Nelli, Rene, *L'erotique des troubadours*, Editions Privat, Paris, 1963.

Osman, Ahmed, *Out of Egypt – The Roots of Christianity Revealed*, Arrow Books, London, 1998.

Pagels, Elaine, *The Gnostic Gospel*, Penguin, London, 1979.

Pennick, Nigel, *Celtic Sacred Landscapes*, Thames & Hudson, London, 1996.

Phillips, Graham, *The Moses Legacy – the Evidence of History*, Pan Macmillan, London, 2002.

Picknett, Lynn & Prince, Clive, *The Stargate Conspiracy*, Little, Brown & Co, London, 1999.

Picknett, Lynn & Prince, Clive, *The Templar Revelation – secret guardians of the true identity of Christ*, Corgi, London, 1988.

Potts, Timothy, *Mesopotamia and the East*, Oxford Univ Press, Oxford, 1994.

Rankin, H D, *Celts and the Classical World*, Croom Helm, London & Sydney, 1987.

Reid, Struan, *The Silk and Spice Routes – Cultures and Civilizations*, Belitha Press, UNESCO, London, 1994.

Rohl, David, *Legend – the Genesis of Civilization*, Random House, London, 1998.

Rudgley, Richard, *Lost Civilizations of the Stone Age*, Arrow Books, London, 1998.

Sadhu, Mouni, *The Tarot*, George Allen & Unwin, London, 1962.

Schoch, Robert M & McNally, Robert Aquinas, *Voices of the Rocks: a Scientist looks at Catastrophes and Ancient Civilizations*, Thorsons, HarperCollins, London, 1999.

Schonfield, Hugh J, *The Essene Odyssey – the Mystery of the True Teacher and the Essene Impact on the shaping of Human Destiny*,

Element, Shaftesbury, Dorset, 1984.

Stern, Sir Nicholas, *The Economics of Climate Change*, Cambridge Univ Press, 2007.

Strachan, Gordon, *Chartres: Sacred Geometry, Sacred Space*, Floris, Edinburgh, 2003.

Sykes, Bryan, *The Seven Daughters of Eve*, Corgi, London, 2001.

Temple, Robert, *The Sirius Mystery*, Arrow, London, 1999.

Temple, Robert, *The Crystal Sun*, Random House, London, 2000.

Trigg, Joseph W, *Origen*, Routledge, New York, 1998.

Tyldesley, Joyce, *The Private Lives of the Pharaohs*, Macmillan, London, 2000.

Van Doren Stern, Philip, *Prehistoric Europe – from Stone Age Man to the Early Greeks*, George Allen & Unwin, London, 1969.

Velikovsky, Immanuel, *Ramses II and his Time*, Sidgwick & Jackson, London, 1978.

Von Eschenbach, Wolfram, *Parzival*, Hatto, A T (tr), Penguin, London, 1980.

West, John Anthony, *Serpent in the Sky – The High Wisdom of Ancient Egypt*, Quest Books, Theosophical Publ House, Wheaton IL, 1993.

White, Michael, *Isaac Newton – the Last Sorcerer*, Fourth Estate, London, 1997.

Whittle, Alasdair, 'The First Farmers', in *Prehistoric Europe – an illustrated History*. ed Barry Cunliffe, Oxford Univ Press, 1998.

Yates, Frances, *The Art of Memory*, Univ Chicago Press, Chicago, 2001.

Notes

1   Andrew Sherratt in particular [Mallory, 1991, p179].

2   Urartian wine was famous – and an earthenware pot found in Miyandoab plain was 'found to contain a dark residue' which was primitive wine sediment dating from possibly the sixth millennium [Rohl, 1998, p146].

3   The parallels between the legend of Jason and Maday Qara are as follows: both fathers of the hero have been usurped; hero brought up by foster parents, one of whom either is a horse or half horse; both heroes have to kill men who are deformed; both heroes warned of dangerous birds about to attack companions; each uses either horse or ship to cross sea and have the back clipped by rocks; arriving in land of usurper find that wife or daughter of usurper is a demon/witch; both have a test involving large animals either camel or bulls; both have to obtain gold coffer or Golden Fleece which determines the fate of the usurper; usurper warned about arrival of hero; usurper dies and wife/daughter attempt to marry hero but are both rejected; rejected demoness/witch kill members of the hero's family (wife, wife's father, children or hero's parents) [Baldick, 2000, p83].

4   The comparison between the Odyssey and Alpamysh is as follows: Alpamysh and his 40 companions tested by 40 slave girls and Odysseus and the Sirens; Alpamysh defeat a Kalmuck army and Odysseus defeats the monster; Alpamysh and companions get drunk and die in a fire which only Alpamysh survives and is thrown into an underground dungeon and Odysseus is the only one to survive the test of eating of the cattle of the Sun; but is rescued many years later by enemy's daughter; hero arrives home just before usurper is about to marry hero's wife; but hero is recognized by his family, although in disguise, and succeeds in an important

test; usurper is then killed and hero resumes rightful throne. Baldick also compares the Homeric epic of the Odyssey with the narrative of the Indian national epic, the *Mahabharata*, in which a usurper takes over the hero's country [Baldick, 2000, pp86-7].

5   The discovery of precession is usually attributed to Hipparchus in 126 BC when he noticed from comparisons with older star charts that the stars had shifted their positions, but it was seemingly a phenomenon known from much earlier times [Hancock & Faiia, 1998, p50; de Santillana & von Dechend, 1977, p59].

6   In March 1998 anthropologist Fred Wendorf announced that he had found a Neolithic ruin at Nabta, west of Abu Simbel, in the southern Sahara. This ruin was on the edge of an ancient lake which had held water from 9,000 BC to 3,000 BC but had dried up in 2,800 BC. What Wendorf found were villages with wells and shrines to cattle, 'a number of flat, tomblike structures', and 'five lines of standing stones'. There was also a stone circle with four sets of stone slabs, some of which were 9ft high: two aligned north to south; and two in a line of sight to the horizon where the summer solstice would have been 6,000 years ago, which, if correct, would make Nabta the 'most ancient astronomical alignment yet discovered' [cited in Schoch & McNally, 1999, pp56-57].

7   Tocharian had two versions, known as Tocharian A and B [Barber, 1999, p118].

8   The same comparison applies to the ancient Persian texts credited to Zarathrustra which are similarly placed in a rural, non-urban, stockbreeding landscape and possibly date from before 1,000 BC [Mallory, 1991, p52].

9   The intrepid Scandinavian adventurer Thor Heyerdahl confirmed the plausibility of Bahrain being on the trade route between Mesopotamia and the Indus valley. Using a seaworthy replica of the craft of that time, in 1977 he made his

voyage from Iraq in an 18m long bitumen-coated reed boat named the *Tigris*. After two days he reached Bahrain and after 50 days he arrived in the Indus valley [Rohl, 1998, p293].

10 The same word for 'merchant' existed in ancient Hebrew ('can'ani' or 'kina'nu') and in Akkadian ('kinahhu') [Aubet, 1993, p7].

11 There are, for example, aristocratic burials at Mycenae on the mainland which date from about 1650 to 1500 BC that include much Minoan material. The destruction of most Minoan sites in about 1450-1430 BC, with the exception of Knossos, could well have been the result of Mycenaean invasion, rather than the volcanic explosion on nearby Santorini [Fitton, 2002, p164, p178].

12 'Phoenix' comes from a Greek word *phoinos*, which refers to the purple color of the bird's plumage and is also at the root of the Greeks' name for the Phoenicians, who were famous for their purple dye [Aubet, 1993, p5].

13 A Helmholtz resonator is a device shaped like a spherical vessel with a smaller opening that 'essentially amplifies those sounds of a given frequency that are part of a complex mixture of sounds, enabling them to be distinguished' [Bakus, 1969, p72].

14 Interestingly, in this context, it is worth noting that alabaster was referred to as *ankh* ('eternal life') and the main quarry in Egypt, from the first half of the third millennium BC, the 4th Dynasty onward, was 25 miles southeast of Tell el Armana and known as *Hatnub* or 'house of gold'. According to Naydler, 'This phrase was also used to designate the sarcophagus chamber in royal tombs' [Naydler, 2005, p164].

15 Hathor, it would appear, had many roles. It was she who 'protects the initiate from hunger and thirst by providing spiritual nourishment in the passage through the *Duat*', the shadow world [Clark, 2000, p157]. Undoubtedly, her horns were certainly a symbol of fertility and nurturing. She even

lent her crown of horns and sun disc to Isis, the goddess most closely associated with Osiris, when Isis was taking on a nurturing role such as being mother to the infant Horus. One can see many examples of the ancient symbol of mother with a child on her knee in the form of terracotta Egyptian models in the Louvre in Paris which predate the Christian image of Mary and Jesus.

[16] Gardner refers to the passages in *Exodus* where Moses burns the golden calf, turns it into powder and then gives it to the Israelites to drink in water [*Exodus* 32:20, quoted in Gardner, 2004, p14].

[17] Even though the existing temple dates from Graeco-Roman times when Egypt was ruled by the Ptolemies, in one of the crypts there is 'an inscription stating that the plan of the temple was established in predynastic times'. Texts from the 18th Dynasty – when Tuthmoses I and III restored the temple – claim that the plan dates from the time of the Followers of Horus, and the alignment of the temple implies that it could be as early as 4000-3000 BC. Certainly, Denderah has been 'regarded as a sacred precinct from the most archaic times' [Clark, 2000, pp221-22].

[18] Not just Uzzah, but two sons of Aaron as well, and others in battle, II *Samuel* 6:1-11, reference the moving of the Ark: 'Uzzah put out his hand to the Ark of God and took hold of it, for the oxen stumbled… and God smote him there because he put forth his hand to the Ark' [*Leviticus* 10:1-2].

[19] What is of interest is that in the Hyksos layer of 1800-1500 BC these deities frequently took the form of an ox, a lion, an eagle and a man. This combination of 'gods' recalls, first of all, Ezekiel's vision of the fiery wheeled object, and secondly, the *Book of Revelation* with John's vision of 'four living creatures' around the throne of God. The simple explanation for this combination is that they were not deities but actually symbols representing four groups of the signs of the zodiac (Aquarius

the Man, Taurus the Bull, Leo the Lion and Scorpio as the Eagle).

20 In addition to Akhenaten's monotheism, other aspects of Egyptian religious practice and belief have informed the Judeo-Christian tradition. The Egyptians had water rituals, for instance, that included baptism. There are images of the baptism of infants, such as the infant Amenhotep III at Luxor. The Ten Commandments are clearly based on the 42 negative confessions that the Egyptian soul was believed to make after death ('I have not committed adultery, I have not defiled the Nile, I have not lied, I have not stolen'…). According to Egyptian writer Ahmed Osman, there are even three elements of the divine aspect of pharaoh that compare with the Jesus story. These elements date from as early as the 4th Dynasty, 27th century BC. They are: the pharaoh's conception and holy birth; his anointment and coronation; and his resurrection. Interesting to note in this context that the word 'messiah' (me-se-he) is of Egyptian origin meaning 'crocodile' and refers to the royal anointment of the pharaoh with the fat of the 'holy crocodile' [Osman, 1998, p283, p274, p275].

21 Ice-core data records a 'significant' acid layer for a period dated 1645 BC (plus or minus 20 years) [Baillie, 1995, p79]; likewise frost ring data describes one frost event in the second millennium BC as being 'severe' and dated around 1626 BC. Irish bog oaks corroborate the frost ring data with samples that include no growth at all after 1626 BC, and there are similar narrow rings in German tree samples for the same period [Baillie, 1995, pp74-75]. Baillie also refers to historical references of the catastrophe in Ireland and China. In Ireland the chaos was so bad that there were seven years without a king. In China, where the effects resulted in summer frosts and failed harvests, Baillie notes that this catastrophe coincided with the start of the Shang Dynasty which saw its chance to gain power [Baillie, 1995, p150].

22 'To your descendants I give this land, from the river of Egypt to the great river, the river Euphrates, the land of the Kenites, the Kenizzites' [*Genesis* 15:18-19].

23 Alan Buttery describes these Sherden as wearing a two horned bronze helmet with a disc on the shaft, and carrying a small round wooden buckler and a large bronze stabbing sword [Buttery, 1974, p24].

24 Steve Jones, the geneticist, considers the Etruscans to be related to the modern Lebanese, who were the Phoenicians [Jones, 2000, p212].

25 There is archaeological evidence which confirms the 'decisive importance' of a Corinthian influence on the Etruscans in the second half of the 7th century BC [Cornell, 1995, p124].

26 Cornell makes the direct comparison between Tusculum and Nazi-controlled France [Cornell, 1995, pp351-352].

27 As Seneca said, regarding the Greek mystery religions, 'Eleusis reserves secrets that it reveals only to those who return to see her. Nature too does not reveal all her mysteries at once... These arcana do not unveil themselves in a hurry, nor to all men. They have been withdrawn to the depths of the sanctuary, well apart in an inner chapel' [Lindsay, 1970, p383].

28 J P Mallory points out that, for example, words for 'cornice, coping, brick' were foreign to both Greek and Latin [Mallory, 1991, p67].

29 Where there is archaeological evidence of megalithic masonry, such as that in the west of the Italian peninsula, it predates the arrival of the Romans. When the Romans invaded the Etruscan territory of Veii in 396 BC they found elaborate drainage tunnels in fields (*curriculi*) and 'carefully engineered roads' [Cornell, 1995, p310]. Likewise, at Aletrium, there are the remains of pre-Roman polygonal walls of the Hernici, a people who joined Rome and the Latin league in 486 BC [Cornell, 1995, p300].

30 Nora Chadwick cites the view of some that 'the occupation of

Britain by the Romans was largely resolved in order to destroy druidism at its roots', quoting in particular Theodore Mommsen, who was writing on the subject in 1909 [Chadwick, 1997, p16].

31 Tacitus also mentions that 'Christus, the founder of the name, had undergone the death penalty in the reign of Tiberius, by sentence of the procurator Pontius Pilate' [Pagels, 1979, pp89-90].

32 Philo of Alexandria (c.30 BC–AD 40) states that they did not offer animal sacrifice [Dupont-Sommer, 1961, p21].

33 The name 'Sampsaean' itself comes from the Aramaic word for Sun, *shimsha*, and is a possible connection with a well known biblical figure who may have been a member of their sect, a man who was famous for not cutting his hair (another Nazirite?), namely Sampson.

34 A copy of the Damascus Document was also found in an ancient synagogue in Cairo at end of the 19th century AD [Baigent & Leigh, 2001, p218].

35 There is evidence in the Habakkuk Commentary that the wicked priest was Ananas [Baigent & Leigh, 2001, p287].

36 The Edict of Milan, 313 AD: 'Freedom of worship ought not to be denied. The right should be given to care for sacred things according to each man's free choice', with the Christian religion first and pagan as 'any other cult' [Frend, 1984, p483].

37 In the opinion of Irenaeus, 'In unbroken succession, each bishop was in accord with the first occupant of his throne, and the first bishop had been appointed directly to it by an apostle (or a disciple of the apostle)' [Brox, 1994, p79, pp86-88]. He attributed his doctrinal authority for such an approach to the New Testament passage in *Acts*, where Peter explains to the others that, since Judas had been allotted a share in 'this ministry', someone else 'must become with us a witness to his resurrection' and so continue the apostleship [*Acts* 1:15-26]. A tradition formed that Peter had been the first

bishop of Rome and therefore enabling the Roman bishops to claim superiority, which they continue to maintain without any historical evidence for this claim.

38 Initially, the term 'papa' had been used for any bishop. It was not until the 6th century Council of Arles that it became exclusively used for the Bishop of Rome when he was given the title the 'most glorious'.

39 The fan, special headgear (the *pallium*), special shoes, the mitre, the gloves, the kissing of the hand and, by the 7th century, the ring.

40 The Essenes believed that, while 'bodies are corruptible, and their matter unstable, souls are immortal and endure forever; that, come from the subtlest ether, they are entwined with the bodies which serve them as prisons, drawn down as they are by some physical spell' – views which laid the foundation for later Gnosticism [Dupont-Sommer, 1961, p33].

41 The Church did not entirely lose the idea of a return in a physical body as it retained the possibility of resurrection at some future time, following a final Judgment Day. Nevertheless, the threat of Purgatory with its emphasis on punishment had the useful effect of encouraging obeisance in the general population.

42 Interestingly, something that we take for granted, the game of chess, probably came to us via the Arabs. The strange term 'checkmate' makes no sense in English but can be explained with reference to two Arabic words *sheik* ('check') meaning 'king' and *ma'at* ('mate'), a clear reference to the goddess Ma'at and her role in the shadow world the Duat, meaning 'death' (a French Algerian friend of mine, Ghali Benamar, translated the French, *echec et mate*, for me into Arabic to confirm this link).

43 The symbolism of the zero has its own significance. Written as a numerical symbol '0', it is not a coincidence that it is the shape of an egg and in this form relates to the Egyptian concept of *bnn* with the meaning 'to beget' or 'embryo'. Thus,

even 'nothing' has potential.

44 The Golden Section later became fundamental to the revival of classical architecture in the Renaissance. This is the number series known as the Fibonacci series in which each number is the sum of the two preceding numbers (1+1=2+1=3+2=5+3=8+5=13, *etc*) and it becomes an irrational ratio when dividing one number by the previous number, resulting in a ratio of approximately 1:1.618.

45 In Baghdad it was a Sufi alchemist, Jabir ibn Hayyam – a man skilled in mathematics, medicine and other sciences, keen on disseminating knowledge of Pythagorean principles of number – who had in his possession one of the oldest copies of the famous Hermetic text called *The Emerald Tablet* and who wrote the magical tales of a *Thousand and One Nights* for caliph Harun al-Rashid [Baigent & Leigh, 1997, p41].

46 In the Sabian version the *magi* visit John and then Herod. Herod feels threatened by their news of John and it is because of him that he instigates the massacre of the Innocents. John's father Zechariah is warned by an angel and flees with John and his wife Elizabeth to Egypt. This alternative view of the birth of Christ from a potentially pre-Christian group could explain the lack of reconciliation between the gospels as it implies that the New Testament version was borrowed from elsewhere.

47 Isaac Newton was, however, able to learn about alchemy and other matters because of a complicated line of transmission through secretive organizations starting with the Templars.

48 This use of the Atbash cipher shows that the Templars came into contact either directly with descendants of the Essenes, or indirectly through the Sabians, because of its similarity to the Pesher code in its use of the Hebrew alphabet. There is evidence of the Essenes' use of the Atbash cipher in the Dead Sea Scrolls found at Qumran [Schonfield, 1984, pp7-9; Andrews & Schellenberger, 1996, p399].

49 'Beyond $E=mc^2$, a first glimpse of a postmodern physics, in which mass, inertia and gravity arise from underlying electromagnetic processes.' [Haisch, Rueda & Puthoff, *The Sciences*, 1994].

BOOKS

O is a symbol of the world, of oneness and unity. In different cultures it also means the "eye," symbolizing knowledge and insight. We aim to publish books that are accessible, constructive and that challenge accepted opinion, both that of academia and the "moral majority."

Our books are available in all good English language bookstores worldwide. If you don't see the book on the shelves ask the bookstore to order it for you, quoting the ISBN number and title. Alternatively you can order online (all major online retail sites carry our titles) or contact the distributor in the relevant country, listed on the copyright page.

See our website **www.o-books.net** for a full list of over 500 titles, growing by 100 a year.

and tune in to myspiritradio.com for our book review radio show, hosted by June-Elleni Laine, where you can listen to the authors discussing their books.

mySpiritRadio